the
SECRET LANGUAGE
of Kittens

the SECRET LANGUAGE of Kittens

THE BODY LANGUAGE OF YOUNG CATS

Tammy Gagne

METRO BOOKS
NEW YORK

METRO BOOKS
New York

An Imprint of Sterling Publishing
387 Park Avenue South
New York, NY 10016

METRO BOOKS and the distinctive Metro Books logo
are trademarks of Sterling Publishing Co., Inc.

© 2014 by Quid Publishing

Editor: Jane Roe
Design: JC Lanaway

ISBN 978-1-4351-5346-2

For information about custom editions, special sales,
and premium and corporate purchases, please contact
Sterling Special Sales at 800-805-5489 or specialsales@
sterlingpublishing.com.

Manufactured in China

2 4 6 8 10 9 7 5 3 1

www.sterlingpublishing.com

CONTENTS

INTRODUCTION

*K*ittens are not cats—not yet. Owners of these young and impressionable animals must not expect their new pets to come home simply knowing what it means to be a cat. They also mustn't expect their young feline companions to behave in the same way that a person, a baby, or even a dog might. It is the owners' responsibility to learn how their kittens express their needs and show their feelings. Only by learning the kitten's secret language can an owner help shape the behavior and forge a positive relationship with this unique young animal.

From the moment you decide to add this tiny animal to your household, you will need to make some big decisions. Certainly, some of these choices will have a greater impact than others. Perhaps you already have a specific breed in mind, or maybe you want to see if any kittens are in need of rescue before selecting your new pet. What matters most is that you have all the facts at your disposal and that you think your decision through completely. Whether you buy or adopt, you never want to contribute to the overwhelming problem of unwanted pets.

Although it may sound complicated, looking at your kitten's experiences from your pet's perspective isn't nearly as difficult as it sounds. Think of it as merely considering your kitten's viewpoint before deciding how to approach a situation. In many ways the feline vantage point is less complicated than a human's. To put it bluntly, people often overthink things, which can make training a kitten a much more complicated task.

Each breed—and each individual animal—is a little different. Certain breeds are known for various characteristics, both physical and temperamental. Kittens also inherit looks and mannerisms from their parents, which can match their breed standard closely or hardly at all.

Mixed breeds are a whole different ballgame, combining the looks and personality traits of numerous different breeds.

Finding the perfect kitten isn't truly about a perfect kitten at all, but rather identifying an animal that is the best match for *your* personality and lifestyle. No matter how gorgeous you think a Persian kitten is, this breed is the worst choice for you if you lack the time or inclination to keep up with grooming this time-intensive breed. Likewise, you may laugh at a Siamese kitten's talkativeness upon meeting the animal, but will you find its nonstop chatter as fun on a daily basis? Only you can make these choices, but it is essential that you give them the time and careful consideration they deserve.

Once you have made your choice, the real fun—and work—begins. Bringing a new kitten home can be a thrilling event that turns into a treasured memory. Owners must have certain items, along with a basic understanding of their new pet's needs, ready before homecoming day arrives. The list of things you will need to buy isn't long, but each item comes with certain choices. Do you buy metal or plastic dishes? Will your pet wear a collar? Where will it sleep? By answering these and other questions before you bring your kitten home, you will make the day a better one for all involved.

Having a new kitten is a bit like having a new baby. This vulnerable animal is depending on you to care for it, feed it, and in some cases entertain it. Here again you will be faced with many choices. For example, a kitten needs to eat food made specifically for younger felines, but which brand or formula is best? How much grooming your kitten needs depends at least somewhat on its breed, but introducing your pet to the grooming process is important during those early days at home whether it has a long or short coat. An animal that is exposed to routine grooming while it is young is much more likely to accept tasks like brushing and bathing as an adult.

While you don't need to know everything about kittens before you bring your new pet home, having a basic understanding of kitten body language and senses can be helpful as you get to know each other. As you spend more time with your kitten, your insight into its secret language will develop far beyond the reaches of this or any other book. Remember, your kitten is an individual. Certainly, it will display many qualities that go along with its

species and breed, but it will also acquire its own ways of doing things and expressing itself. And you will become impressively in tune with your pet's personality as time passes.

When a health problem occurs, an owner is often the first to notice the issue. Sometimes the warning comes from reading up on your kitten's breed, so you know the common ailments to which your animal is prone. Other times it's not a specific symptom that alerts an owner, but rather a general feeling of unease. Owners who know their kittens best know when something isn't right. Even if you can't articulate the problem, you may know that it's time to see the veterinarian.

As you care for and train your new pet, you may not notice how fast time is passing. It is while you are taking care of the more mundane business, such as litter-box training and taking your kitten for booster shots, that you will bond with your pet. Of course, more pleasing activities like playtime and napping together on the sofa are also ways to deepen your relationship with your kitten, but it is actually a combination of the fun and ordinary tasks that will strengthen the trust and bond between you the most.

One of the things that makes kittenhood so special is that this period in your pet's life is so fleeting. Before you know it, your tiny kitten will be on the verge of young adulthood. Like human parents, kitten owners often exclaim with sincere wonder, "Where did the time go?" As you enter this new phase of your pet's life, your young cat will continue to go through changes, but you will be better prepared for all of them, because you took the time to learn the secret language of kittens.

No. 1 OWNING A KITTEN

RESPONSIBILITIES AND REALITIES

Just one look at a newborn kitten and many people become instantly hooked. In many cases the decision to buy or adopt a young cat marks the beginning of a long and rewarding interspecies relationship. You will almost certainly get the same amount of love back—or even more—as you put into caring for your furry new charge. But it is important to understand that you are making a long-term commitment by becoming a cat owner.

LONGEVITY

One of the best traits of the feline species is its lengthy lifespan. Most cat breeds can live as many as 20 years or more, a fact that many a dog owner envies. A cat's longevity, however, also means you must be certain that owning one of these animals is right for you. Choosing the best kitten for your lifestyle and personality is vital before entering into this long-term relationship. The various cat breeds and mixes can range in temperament greatly, as can individuals of the same breed. The best way to be sure that you have found your ideal match is by spending some time getting to know a kitten before you take it home.

It is also important to understand that your little ball of fur is going to do a lot of growing and changing during its first year or two of life. Most kittens do not change significantly as they move into adulthood, but a certain amount of physical and temperamental change should be expected. Fortunately, changes that come with maturity are usually welcome ones. For example, an energetic kitten often mellows a bit by the time it reaches the age of two or three. In some cases, though, a cuddly kitten might become more independent as an adult. You must be willing to accept your kitten's adult personality, even though it will take some time to reveal itself.

BEHIND THE SIGNS

Are Two Kittens Better Than One?

One way to help prevent boredom, and subsequently unpleasant behaviors, is to get two kittens at once. When cats are raised together, they tend to be more social and well adjusted. Some owners worry that having two young cats at the same time might make tasks like litter-box training more difficult, but kittens learn largely by example. Whatever you successfully train one kitten to do, the other is likely to copy. Having two kittens can be a lot of fun, but it is essential that you truly want both kittens before you decide to double up. Each and every animal deserves to be wanted and loved.

GIVING IT TIME

Spending time with your kitten regularly will help you to forge a strong bond. Some adult cats can be extremely autonomous, but most still enjoy spending time with their human companions. This one-on-one time is especially important for younger felines.

Make a point of petting your kitten whenever the opportunity presents itself. Also, make time for at least one or two play sessions each day. How much time you devote to waving a toy fishing pole with a feather for bait around is far less than important than making the time to do it—or whatever your kitten enjoys most. As you get to know your pet more, you will discern which pastimes it enjoys.

Once you have cat-proofed your home, it will be safe to leave your new kitten alone for reasonable periods, but it is wise to spend as much time with your pet as possible in those first few days and weeks. If you work long hours, consider scheduling your vacation time around your kitten's arrival. In addition to bonding, this time gives you the opportunity to litter-box train your new pet if necessary. A kitten that becomes housetrained early is usually more reliable with this task as an adult. Spending time with your new pet is also a great way to identify areas in which you may need to tweak the kitten-proofing.

If you cannot take time off from work or your other responsibilities, consider enlisting help. Some cat breeds are known for bonding most closely with a single person, but many breeds can form positive relationships with two or more household members. A spouse, older children, or extended family members might be able to spend time with your kitten when you cannot.

PAYING THE BILLS

Before you become the proud owner of a kitten, you must consider whether you can afford the financial responsibilities that come with your new household member. You will need a few basic items before bringing your kitten home. Many of these supplies are one-time purchases, such as feeding dishes and a comfortable bed. Others will be repeated expenses: Nutritious food, regular veterinary care, and cat litter (for indoor pets) are just a few of the things you will need to buy for your pet throughout its life.

Preventive veterinary care is the best way to ensure that your kitten remains in good health. But even healthy cats get sick from time to time. In this situation you will need to make additional trips to the vet. You must be prepared to handle the extra costs of sick visits, medications, and even emergency care when necessary.

NOT ALL FUN AND GAMES

Some people mistakenly assume that cats are low-maintenance pets. Although it's true that cats generally require less supervision, training, and grooming than many dogs, owning a cat is not without its day-to-day responsibilities. Some cat breeds even require more care than dogs or other household pets. If you want a Persian kitten, for instance, be ready to perform daily brushings to keep its fur clean and free of tangles.

You are unlikely to come home to find that your new kitten has chewed up your beloved Manolo Blahniks, but your precious pet may get into mischief in other, equally expensive ways. Most cats, including young kittens, enjoy scratching. If left to its own devices, your kitten may decide that your living room sofa is the perfect object for this activity. Luckily, most owners can redirect their cats to use a scratching post instead.

It should also be noted that some kittens do indeed chew their owners' belongings. Like puppies, cats go through a teething phase that typically begins around four months of age. This painful stage can last until your pet is eight months old. If your kitten is a chewer, you must make a point of putting your shoes (and anything else you don't want ruined) in a secure spot. If your kitten has a penchant for electrical cords, this could prove to be one of those areas in which additional kitten-proofing of your household is needed.

IN OR OUT?

The argument about whether cats should be allowed to roam freely outdoors has been going on for decades. At one time nearly all pet cats went outside to relieve themselves, hunt, and get exercise. Many still do. But many owners have come to realize that having a free-roaming cat comes with certain risks. The biggest of these are automobiles, wild animals, and diseases spread by feral cats. The effects of all three can be deadly.

Many owners worry that an indoor cat will become bored over time. Others think that being kept indoors all the time is what drives some cats to destructive behaviors like inappropriate chewing and scratching. Some also add that indoor cats are more likely to become overweight from lack of activity. The good news is that you can prevent all of these problems while still keeping your pet safe. Interactive toys can provide both mental and physical stimulation. They can also serve as excellent distractions from unwanted behaviors.

At the same time, it is true that some cats have a stronger urge to explore the great outdoors than others. Older cats that have been outdoor cats in the past, for instance, may have a hard time adjusting to being indoor cats. But kittens can usually be raised to become content, indoor pets.

Bear in mind that indoor cats don't have to stay indoors all the time. Getting fresh air and exercise is good for everyone—human, feline, and otherwise. The trick to offering your kitten a safe outdoor experience is using a leash or an outdoor enclosure. Many cats can be taught to walk on leashes. Kittens are often most open to this practice. Any habit that you begin while your cat is young is more likely to be tolerated well. Some breeds in particular, though, take to leash walking especially well. The Abyssinian, for example, has high aptitude for leash training.

Outdoor enclosures are another viable option if you want to provide your kitten with regular outdoor time. From a standard wire pen you can purchase at your local pet supply store to a one-of-a-kind structure you design and build yourself, your kitten's enclosure can be as

 Kitten Fact

If you allow your kitten to wander outside, affix an identification tag to its collar. It is also smart to have an outdoor pet microchipped. Nearly three-quarters of all lost cats and dogs with microchips are returned to their owners. Even if your kitten will be an indoor pet, having it microchipped could reunite the two of you if your kitten ever manages to get out.

simple or elaborate as you like. What matters most is that it is secure. Your kitten should not be able to detach or slip through the wires.

Whether you use a leash or an enclosure, it is also important to pay close attention to the weather. Outdoor cats that are used to the elements may tolerate them fairly well, but your young kitten may become chilled easily. Your first responsibility as a pet owner is to keep your kitten safe at all times.

No. 2 THE RIGHT KITTEN FOR YOU

PEDIGREED, MIXED-BREED, RESCUE

*B*efore you can start searching for your kitten, you must decide if you want a pedigreed cat or a mixed-breed animal. Each choice has clear benefits. Which one is right for you is mostly a matter of personal preference, but knowing the advantages of each can save you time and energy—and in some cases even money.

WHAT'S IN A PEDIGREE?

Pedigreed cats are purebred animals. Owners can trace the lineage of these kittens through numerous generations with all the ancestors belonging to the same breed. Just like dogs, cats are available in a wide variety of breeds. The Cat Fanciers Association (CFA) recognizes 40 official breeds. Some, like the Egyptian Mau, have been around for centuries. This breed can be traced back to about 1550 BCE. Others, like the Savannah, are much newer. This breed was accepted by The International Cat Association (TICA) in 2001, and was awarded championship status in 2012.

Each breed has its own characteristic appearance and temperament. Breed clubs use this unique list of traits to create official breed standards. The best breeders take great care in selecting healthy cats that most closely match this description for their breeding programs. When selecting a purebred kitten, it is important to meet one of its parents, if possible. Seeing the mother (dam) and father (sire) will give you the best idea of how your kitten will look and act as an adult cat down the road.

Consistency is the overwhelming benefit to choosing a purebred kitten. If you want a kitten with lots of energy and a short coat, you can narrow your choices by looking for a breed that fits this

description, such as the Oriental Shorthair. In general, purebred cats have more health problems than mixed breeds do. Over time, inbreeding has caused different breeds to be prone to certain illnesses. The Devon Rex, for example, is prone to hip dysplasia.

You can lessen your chances of selecting a kitten that will develop this and other diseases by asking the breeder whether the breeding pair has been tested for the health issues most common in that breed. A kitten that develops this problem will most likely have inherited it from one or both of its parents. Choosing a kitten from parents that have not developed the problem isn't a guarantee that your pet won't someday suffer from it, but it definitely stacks the odds in its favor.

If you want to participate in confirmation shows with your kitten, you might want to limit your search to purebred kittens. Confirmation, the official term for showing, is a pastime in which judges rate cats by how well they match their individual breed standards. A kitten whose parents have already earned titles in the show ring has the best chance of excelling there itself. You can still take part in showing even if you go with a mixed breed, however. The TICA offers household pet classes in which non-purebred cats are judged on their condition, beauty, and show presence.

 Kitten Fact

Cats suffer from more than 250 different hereditary disorders. Many are similar to genetic illnesses that affect people. Both felines and humans can suffer from retinitis pigmentosa, a degenerative eye disease that causes blindness. In kittens, this illness has also been linked to the deficiency of an amino acid called taurine.

Kittens can start competing in confirmation shows as early as four months of age. A kitten competing in the household pet class must be spayed or neutered once it reaches eight months. No such requirement exists for purebreds, as technically confirmation is a means of evaluating breeding stock.

The best breeders often have an excellent eye for predicting which young kittens will grow into the best show cats. It is important to note, however, that no one can guarantee that a specific animal will win awards. Beware of any breeder who offers such promises. Most importantly, choose a kitten that you find cute and friendly. The most important judge of these qualities for your kitten, after all, is you.

MIXING IT UP

One of the best ways to prevent hereditary problems is keeping a kitten's gene pool as diversified as possible. For this reason many cat lovers opt for a crossbred kitten. Crossbreeds are created by mating two different breeds, with the resulting litter inheriting a wider range of characteristics. Many hybrids, as these kittens are called, will not be eligible for showing. Breed registries such as the CFA and TICA only recognize certain crosses that have developed consistency of both appearance and temperament. If you do not plan to show your kitten, though, this limitation won't affect you.

Hybrid kittens may have two domestic parents or one domestic parent in addition to a wild parent. The Bengal is one of the most popular hybrids, created by crossing domestic breeds with a wild species—in this case, the Asian Leopard Cat. The Tonkinese, however, was created by crossing two domestic breeds, the Siamese and the Burmese.

Intentional hybrids aren't your only option for finding a healthy, mixed-breed kitten. You may also consider taking home a kitten with an unknown pedigree. A kitten such as this is often referred to as simply a domestic shorthair, domestic longhair, and in some cases even a domestic medium hair. While many of these mixed-breed litters are not planned—and therefore do not include carefully chosen parents—they are often healthier than many purebred kittens.

🐾 *Kitten Fact*

A litter of kittens is also called a kindle. The average litter usually consists of two to five kittens.

COMING TO THEIR RESCUE

A breeder isn't the only resource for a healthy, friendly kitten. If you want to make a big difference in the life of a young feline, consider adoption. Many widespread misconceptions exist about adopting animals. One of the biggest is that only older pets are available through shelters and rescue organizations. However, the average age of animals surrendered to shelters in the United States is just 18 months. Many animals are even younger.

When owners of pregnant cats have trouble finding homes for the kittens, they often surrender the animals to rescue organizations or shelters. For this reason, numerous litters—and sometimes the expectant mothers as well—are available through adoption. The best way to prevent this problem is by spaying and neutering most pet cats, but since so many owners fail to have their pets sterilized, the unwanted cat population is growing. From 2010 to 2012, the number of cats surrendered to the Royal Society for the Prevention of Cruelty to Animals (RSPCA) in Britain rose from 29,269 to 31,556—nearly 8 percent. To make matters worse, the number of adoptive owners dropped by 10 percent during this same time period.

It isn't just mixed-breed kittens that find themselves in need of homes. Purebred kittens are also available for adoption through many rescue facilities. In many cases it is a breed's overwhelming popularity that causes this problem. People who jump into cat ownership by purchasing a kitten without thinking it through completely are often unprepared for the responsibility. Many of the kittens obtained this way end up being surrendered when their owners realize

that being a pet owner is tougher than they expected it would be. Some young cats, like Himalayans, are given up because of their intense grooming needs. Others, like the Siamese, are surrendered when their owners realize how vocal they are.

By choosing to adopt a kitten instead of buying one, you can save a life. You will likely receive a wonderful companion in the process. As long as you choose your kitten carefully, making sure that it is the right breed or mixed-breed for you, the kitten you adopt can make as excellent a pet as any animal you might buy from a breeder instead.

Kitten Fact

The RSPCA estimates that 85 percent of cat litters are unplanned.

Fostering

Fostering is a great way to try out the kitten-owning experience and to learn what type of kitten is right for you. The best way to ensure that a homeless litter receives proper care is by placing both the mother and kittens in a loving foster home until the kittens are weaned. Kittens are more likely to grow into friendly cats if they are handled frequently during their first seven weeks.

Contact your local pet shelter or rescue organization to find out what is involved. Many organizations offer special training classes to prepare potential foster owners for the experience. If you form a particularly strong bond with one of the animals, you might even end up keeping it.

Less Than Perfect

Some kittens arrive in shelters with unpleasant conditions—from severely matted coats or fleas to malnutrition or chronic illnesses. A longhaired kitten with shaved fur might not be the prettiest animal in the shelter, but its chances of being adopted are probably far lower than a perfect-looking kitty. And its hair will almost certainly grow back. Many chronic health conditions, such as Feline Leukemia Virus (FeLV) and Feline Immunodeficiency Virus (FIV), can be managed with medication. When given a second chance, many kittens with these and other challenges can make wonderful pets.

CAN A FERAL KITTEN BE TAMED?

Just a single stray cat can give birth to numerous litters during its lifetime. In addition to being homeless, all these kittens also face another stumbling block. With no human interaction, they lack the social skills that make a kitten a pleasant pet. Known as feral kittens, these animals can present added challenges for even the most seasoned cat owners. But is a feral kitten beyond hope? Not necessarily, but experts don't offer rescuers much hope either.

Jesse Oldham is the American Society for the Prevention of Cruelty to Animals (ASPCA) Senior Administrative Director of Community Outreach. He is also the founder of Slope Street Cats, an organization specializing in feral cat welfare. Even with his extensive experience, he hasn't had much luck domesticating these animals. "I, like many first-time rescuers, tried to socialize a feral cat," he states. "He remained under my bed for over a year before I could even touch him. With so many adoptable domestic cats and kittens who are truly happy being indoors, socializing a feral cat should not be the goal."

No. 3 KITTEN BREEDS

UNDERSTAND CHARACTER TRAITS AND CARE REQUIREMENTS

*D*eciding whether you want a purebred kitten or a mixed-breed is only the first step in choosing your new feline companion. If you go with the former choice, you must next decide which breed suits you best. While the CFA only recognizes 40 official cat breeds, you may choose from nearly 100 different breeds of kitten in all. Some breeds look remarkably alike, but each one offers its own unique mix of characteristics such as physical appearance, temperament, and activity level. Some breeds are downright unique in every way.

There is no such thing as the perfect cat breed for everyone. Just like animals, each person is different. The traits that top your list of must-have qualities may rank at the bottom of someone else's. Your goal is to find the perfect kitten for both your personality and lifestyle. And breed is an excellent starting point for narrowing your choices.

SIZING THINGS UP

No matter which breed of kitten you choose, your tiny new pet won't stay tiny for long. Depending on the breed you select, however, your kitten may grow a little or a lot over the next few months. To many people it may seem like all pet cats are essentially the same size once they reach adulthood. They certainly don't vary in size as much as dogs do, for instance, but you may be surprised at just how big or small a particular cat breed actually is. Some large breeds rival small dogs in terms of their size. Other cats could be mistaken for kittens even as adults.

Weighing between 4 and 7 pounds (1.8 and 3.2 kilograms), the Singapura usually wins the title of smallest breed. Since individuals vary, however, a Singapura might be slightly larger or smaller than another diminutive breed. The Munchkin stands out not only for its overall compact size—it weighs between 5 and 9 pounds (2.3 and 4.1 kilograms)—but also for its unusually short legs. Due to a genetic mutation, this breed's legs are about half the length of other cats' legs. The shortness in no way hinders the kitten's ability to run, jump, and play, however. The Cornish Rex, Devon Rex, and German Rex are also among the smallest breeds available today.

On the other end of the size spectrum, we find several other distinct breeds.

Typically ranked the largest cat breed, the Savannah is often described as huge. This breed can weigh as much as 25 pounds (11.3 kilograms). Smaller Savannahs, though, may weigh as little as 10 pounds (4.5 kilograms). Most Savannahs have the long-legged look of the wild African servals in their pedigrees. Other kittens that will grow into the largest adult cats include the Maine Coon, Norwegian Forest Cat, and the Ragdoll.

If a small cat is too tiny and a large cat is too big, a medium-sized cat may be, as they say in fairy tales, *just right*. And if you are searching for a kitten that will grow into a medium-sized cat, you have plenty of choices. From the American Curl to the York Chocolate—and 28 other breeds in between—you are sure to find a breed to your liking. With so many choices in fact, it may be time to start considering the other ways these and other breeds differ.

THE LONG AND THE SHORT OF IT

For many people searching for a new kitten, one of the most important factors is coat length. Before you make up your mind between a longhaired and shorthaired breed, however, be sure you have all the facts. Many people are surprised to learn that shorthaired breeds shed just as much as those with longer fur. A Bombay may leave just as much fur behind on the sofa as a Turkish Angora. Even the Sphynx, which is largely hairless, will lose the small amount of hair it has over time. New hairs immediately replace the old ones, essentially pushing the older hairs out and starting the cycle all over again.

You can reduce the amount of hair your kitten leaves behind on your clothes and furniture by brushing your pet regularly. If you go with a longhaired breed, however, brushing will almost certainly be a mandatory task, not an optional one. Breeds like the Chantilly (also known as the Tiffany), for instance, require daily brushing.

Longhaired breeds are undeniably beautiful, but they aren't for everyone. Regular brushing removes dead hair, dirt, and other debris from a kitten's coat. In the case of a longhaired breed, brushing also prevents mats from forming. If you don't find the idea of brushing a kitten every day discouraging, you must still be sure you have the time to dedicate to a longhaired cat's grooming needs. In addition to matted fur, a longhaired cat can develop a severe problem with hairballs if its owner drops the ball on this essential task.

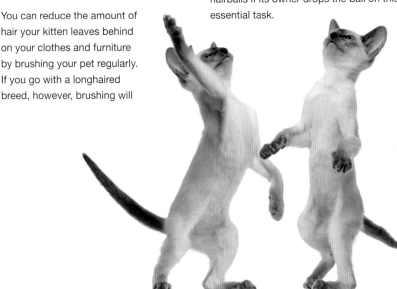

If you or someone in your household is allergic to cats, you might think the problem is the cat's hair. The truth is that most allergies to cats are triggered by cat dander. Like its hair, your kitten's skin is constantly growing. The outermost skin layer, called the epidermis, sheds much like its fur. As the new layer rises to the top, the dead skin flakes off. For this reason getting a shorthaired cat will not be a solution to an allergy problem.

Thankfully, some cat breeds are considered hypoallergenic, meaning that they produce significantly less dander than other cats. The prefix *hypo-* means low, so there is still a chance that an allergy sufferer won't be able to tolerate a hypoallergenic breed. The best way to tell is by spending time with a specific kitten before making the decision to take it home.

BEHIND THE SIGNS

Hypoallergenic Kitten Breeds

The following breeds are known for their hypoallergenic properties, although some people may still experience allergic reactions.

- 🐾 **Balinese**
- 🐾 **Bengal**
- 🐾 **Cornish Rex**
- 🐾 **Devon Rex**
- 🐾 **Javanese**
- 🐾 **LaPerm**
- 🐾 **Oriental Shorthair**
- 🐾 **Russian Blue**
- 🐾 **Siberian Forest Cat**
- 🐾 **Sphynx**

🐾 Kitten Fact

The Selkirk Rex is available in either a long- or shorthaired variety. Both can have curly coats, although this trait is most easily seen in longhaired kittens. If you want your Selkirk Rex kitten to keep its curly coat, you mustn't brush it as often as other breeds. Too much brushing can straighten the hair, leaving it frizzy.

PURR-SONALITY

While it is easy to notice a kitten's physical attributes first, its other less tangible traits are more important in many ways. Physical beauty makes a kitten pleasing to the eye, but the right temperament is what will make it an enjoyable pet. The prettiest cat can be the loudest and most annoying. A toddler may enjoy petting a longhaired kitten's soft coat, but he definitely won't find being scratched by the same animal very pleasant if it isn't tolerant of young children. And a person who has to work early each morning won't appreciate a kitten that spends all night wreaking havoc around the house—no matter how cute it is.

Of course, may owners prefer vocal kittens to quiet ones. Likewise, breeds that aren't especially good with kids can make excellent pets for adults. And an active person may enjoy the companionship of a kitten that is equally energetic. It's all truly a matter of personality—both human and feline.

If you want a kitten that enjoys sleeping at night, consider a Russian Blue. This breed may need a little time to get comfortable in its new surroundings, but once it does, you will most likely find it curled up on your bed beside you whenever you sleep. While many kittens are curious, Ragdolls are unlikely to get into trouble around the house at night or during the day. This breed prefers following its favorite person from room to room rather than exploring on its own.

If you desire a more active pet, a Somali kitten may be best for you. This breed is both active and intelligent, making it an ideal choice for an owner interested in trying something new, like agility training. Yes, some cats can be trained to jump through hoops, literally. A Somali enjoys being the center of attention, doing best as the only cat in the household.

The Burmese is extremely friendly. This kitten will be quick to curl up on its owner's lap, but it will then proceed to *talk* at length. Unlike the Somali, a Burmese is actually better off with other animals in the household, especially if its owner works long hours. If it doesn't have another cat or dog to keep it company, it will likely find some trouble to get into when it is left alone.

Siamese kittens have a reputation for being especially good with kids. Their strong curiosity drives them to insinuate themselves into all sorts of household activities. This breed is often described as dog-like. Many Siamese kittens even enjoy being walked outdoors on leashes.

The exact number of registrations differs each year, but the Persian and the Siamese typically top the list of most popular cat breeds.

THIS BREED IS KNOWN FOR ...

Breed		Breed	
Abyssinian	🐾 🐾 🐾	Kashmir Cat	🐾
American Bobtail	🐾 🐾	Korat	🐾 🐾 🐾
American Curl	🐾 🐾 🐾	Maine Coon	🐾 🐾 🐾
American Domestic	🐾 🐾 🐾	Manx	🐾 🐾 🐾
American Shorthair	🐾 🐾	Munchkin	🐾 🐾 🐾
American Wirehair	🐾 🐾	Nebelung	🐾 🐾
Balinese	🐾 🐾 🐾	Norwegian Forest Cat	🐾 🐾 🐾
Bengal	🐾 🐾	Ocicat	🐾 🐾 🐾
Birman	🐾 🐾 🐾	Oriental Shorthair	🐾 🐾 🐾
Bombay	🐾 🐾 🐾	Peke-Faced Cat	🐾
British Shorthair	🐾 🐾	Persian	🐾
Burmese	🐾 🐾	Ragdoll/Ragamuffin	🐾 🐾 🐾
Burmilla	🐾 🐾 🐾	Russian Blue	🐾 🐾 🐾
California Spangled Cat	🐾 🐾 🐾	Savannah	🐾 🐾 🐾
Chantilly/Tiffany	🐾 🐾	Scottish Fold	🐾 🐾 🐾
Chartreux	🐾 🐾 🐾	Selkirk Rex	🐾 🐾 🐾
Colorpoint Shorthair	🐾 🐾 🐾	Siamese	🐾 🐾 🐾
Cornish Rex	🐾 🐾 🐾	Siberian Forest Cat	🐾 🐾 🐾
Coupari	🐾 🐾	Singapura	🐾 🐾 🐾
Cymric	🐾 🐾 🐾	Snowshoe	🐾 🐾 🐾
Devon Rex	🐾 🐾	Sokoke Forest Cat	🐾 🐾 🐾
Egyptian Mau	🐾 🐾	Somali	🐾 🐾 🐾
Exotic Shorthair	🐾	Sphynx	🐾 🐾 🐾
German Rex	🐾 🐾 🐾	Tonkinese	🐾 🐾 🐾
Havana Brown	🐾 🐾 🐾	Turkish Angora	🐾 🐾 🐾
Himalayan	🐾	Turkish Van	🐾 🐾 🐾
Japanese Bobtail	🐾 🐾 🐾	York Chocolate	🐾 🐾 🐾
Javanese	🐾 🐾 🐾		

 High Energy Level   **Talkative Nature**   **Rapport with Children**

4

SELECTING YOUR KITTEN

A GUIDE TO BUYING OR ADOPTING

*I*f you already know the qualities you want in your kitten, finding your new pet will be much easier. Being able to narrow your search to shorthaired kittens with outgoing personalities, for example, can save you a great deal of time. Many shelters allow prospective adoptive owners to search online profiles based on this kind of information. Breeders have a special advantage in that they have been able to spend a great deal of time with both the kittens and the young cats' parents, allowing them to identify the kittens that best match your desired traits.

But how do you choose between several different kittens that all seem to be perfect for you? Putting just a little more time and effort into the selection process can often help in this situation. First, if you haven't met the kitten in person, arrange for a time to do so. Viewing photos or videos online may be helpful for finding available kittens, but nothing can replace a face-to-face meeting. Pay special attention during your visit, as several details can help you determine which kitten is best for you.

HELLO, BRIGHT EYES

Whether a kitten comes from a breeder or a shelter, the same basic qualities indicate a healthy animal. The most noticeable of these are bright, clear eyes, clean ears, and a fluffy, mat-free coat. Red eyes are a sign of conjunctivitis. Cloudiness can signal a corneal problem or inner eye disorder. Similarly, eye or nasal discharge may be a sign of a respiratory infection. A kitten's nose should be cool and damp, but not running.

A kitten's ears should be clean with no offensive odor. A foul smell or dark-colored ear wax is a sign of mites. The presence of mites—or fleas—does not necessarily mean that a kitten is unhealthy, but it does suggest that the animal isn't receiving optimal care. Most shelters will make sure that problems like these are addressed before clearing a kitten for adoption. Mites or fleas can strike any animal, but a good breeder will take steps to protect kittens from these insidious external parasites.

Bald patches within a kitten's coat could be a sign of mange or ringworm. There is nothing wrong with adopting a kitten dealing with this or another health problem, but you should know about any ailments upfront. Most illnesses can be treated or managed with the help of a veterinarian. Certain medications and treatments can be expensive, though, so you may need to figure them into your budget.

Many kittens with white coats and blue eyes are deaf. This congenital condition will not keep a kitten from making a wonderful pet, but it will change the way you interact with your animal. Again, you simply want all the facts before making your decision.

If a kitten has a bulging tummy, there are two likely causes. One possibility is that the animal hasn't been receiving an adequate amount of food. The stomach responds to starvation by swelling. Fortunately, this problem can be fixed easily—simply by providing the kitten with nutritious food at regular intervals. It's important that you take the animal to a veterinarian, though, to rule out other problems. If lack of food isn't the culprit, worms might be. This health issue can also be dealt with relatively easily in most cases, but it is vital to seek veterinary treatment promptly.

PLEASED TO MEET YOU

Kittens quickly develop a rapport with people they like. If a breeder has several kittens available, the right one for you may very well be the one that comes to you first—especially if you want a highly sociable pet. If this is the case, sit back and wait to see which kitten makes that all-important first move. Also, pay attention to the kitten's eyes. A friendly, curious animal will look at you with its eyes wide open.

If you want a playful pet, take a table-tennis ball with you to your first meeting. A playful kitten won't be able to resist chasing this toy. You can also find similar balls at your pet supply store, some with noisemakers inside that just make them all the more fun for young, energetic cats. If the kitten you are considering has no interest in the ball when you roll it across the floor, the lack of reaction could be a sign that this animal has a more subdued temperament.

Another great way to get a feel for a kitten's personality is watching the animal interact with its siblings. If the kitten initiates play with its brothers and sisters, it will likely behave similarly with human family members in its new home. A kitten that tends to hang back and watch its siblings play instead of jumping into the fray, on the other hand, is more likely to have a more bashful disposition. Neither type is better than the other, but one type may be better suited to you.

If you have younger children, a playful cat is definitely a better option. Any kitten that will be sharing its home with kids needs to be tolerant of being handled frequently. Children should always be gentle when handling any animal, but kittens too should respond to being picked up in a good-natured manner. Wriggling is perfectly normal; biting, hissing, or scratching is not. These are all signs of a poor temperament. None of them mean that the kitten cannot become a well-behaved cat in the future, but it is a sign that this particular animal may be better off in a more mature household.

FIRST FRIENDS

Kittens learn a great deal of their social skills while playing with their littermates. From understanding that they can hurt others when they bite or scratch to realizing that the others will bite and scratch back, this time with its brothers and sisters will help your kitten become a better companion for you. It is with its siblings too that your kitten will learn how to bond with other living beings. For this important reason, it is wise to let your new pet stay with its biological family until about 12 weeks of age. It is between 10 and 12 weeks that the majority of this instructive playtime takes place. Kittens simply do not learn these lessons as well from human family members.

Kitten Fact

Most kittens actively play with their siblings until they are 12 weeks old. By 14 weeks, though, objects like toys will start attracting more of their attention.

QUESTIONS AND ANSWERS

Don't be afraid to ask questions. Jot down everything you want to know before your visit. Listen carefully to the breeder's or shelter worker's answers.

If you are dealing with a breeder, ask which vaccinations the kitten has already had. A written copy of this information should be provided along with the sales agreement, TICA registration paperwork, and your kitten's pedigree. The latter two only apply if the kitten is a purebred animal. You may also want to confirm the kitten's age and make sure that it is old enough to leave its mother. A breeder may allow you to meet a kitten before it is fully weaned, but you probably won't be able to take it home until it is 12 weeks old. No kitten should leave its mother unless it is *at least* eight weeks of age.

Ask if the breeder offers any health warranty. This too should be put in writing. Good breeders guarantee that their kittens are free from Feline Leukemia Virus (FeLV) and Feline Immunodeficiency Virus (FIV). The best breeders urge you to take your new kitten for a routine checkup within the first few days after taking it home to confirm that it is in good health. If the kitten is suffering from a major health problem, the breeder should be willing to take the kitten back and refund your money or give you another kitten when one becomes available.

Responsible breeders will also have questions for *you*. It is their responsibility to make sure that their kittens go to good homes with owners who are both willing and capable of taking care of the animals. You must have the time and financial resources to provide it with everything it needs.

A breeder may ask you to sign an agreement to have your kitten spayed or neutered by a certain age. This will keep your kitten from adding to the unwanted animal population and will also reduce its chances of suffering from certain types of cancer—and eliminate its risk of suffering from other types.

Owners surrendering animals to a shelter do not always volunteer information about the animal's health. For this reason, shelters often perform temperament testing and medical examinations before placing animals in a new home. If the kitten is old enough, it may also be spayed or neutered.

An animal shelter will want to know a little bit about you before allowing you to adopt one of its kittens. In most cases the application process is relatively simple. The workers will want to know if you have any experience as a cat owner, whether you have any other animals, and the number and ages of the other people living in your household. A shelter may also ask for written permission from your landlord to own a pet cat if you live in an apartment. The last thing that the shelter wants is for your kitten to be surrendered again. The workers' top priority is finding each animal a loving, permanent home.

Shelters rely heavily on donations and most charge a nominal adoption fee. This money goes to providing the animals with food and medical care. If you can afford it, consider adding at least a small amount to this fee as a donation. It will allow the shelter to help other animals that aren't as lucky as the kitten you decide to take home with you.

BEHIND THE SIGNS

Questions to Ask a Breeder

🐾 At what age do you allow kittens to go home with their new owners?

🐾 Do you offer a health warranty? What does it cover?

🐾 Do you require a spay/neuter agreement?

🐾 What other paperwork will I receive with my kitten?

🐾 What brand and formula of food is my kitten beign fed?

🐾 Can you recommend a veterinarian?

🐾 Do you offer any follow-up support for new owners?

5

KITTEN-PROOFING
YOUR HOME

PREVENTION AND SAFETY

efore you bring your new kitten home, you may want to complete the essential task of kitten-proofing. Each area of your home to which your pet will have access needs to be considered, in many cases adjusted, and then deemed safe for your young cat. Kittens are even more curious than adult cats. They can get into trouble remarkably quickly and in some surprising ways. While some problems can be minor ones, others can be deadly. Skipping the kitten-proofing process simply isn't worth the risks.

MAKING A START

If you have already brought your kitten home, don't despair. You can begin by kitten-proofing one room and confining your new pet to that single area until you have finished going over the rest of your residence. Actually, limiting your kitten to just a single room in the first 24 to 48 hours is a smart idea, as it allows your pet to acclimate to its new environment slowly. When given access to an entire household, some kittens can become overwhelmed or even frightened by all the new sights and sounds.

If you do have some lead time, begin with the room in which your kitten will spend most of its time. If you are unsure of where this might be, think of where you will place important items like your pet's dishes and litter box, as these rooms will likely be frequented the most by your new four-legged friend. Your kitten should have access to fresh water and a place to relieve itself at all times.

In order to kitten-proof your home properly, you must start thinking a bit like a kitten. Anything shiny, crinkly, or otherwise entertaining to the feline

species is fair game to a kitten in search of a new toy. As silly as it may sound, it can even be helpful for you to get down on the floor to look at a room to get a true feel for your tiny pet's perspective. In doing so you just might discover a potentially dangerous object or two that you may have missed otherwise.

 Kitten Fact

Many pet experts think that collars are dangerous for cats and kittens, but a 2013 study by the Universities Federation for Animal Welfare found little data to back up this theory. The study stated, "Interviews with 107 veterinarians indicated an average rate of one collar injury observed per 2.3 years of veterinary practice." Still, many owners skip the collar to err on the side of caution. Owners who do choose to use a collar can reduce the risk of injury by purchasing a collar with breakaway technology. This smart item is designed to release quickly in the event that it becomes caught.

EVERYTHING IN ITS PLACE

You can check several tasks off your kitten-proofing to-do list by simply doing some light housekeeping. Picking up and putting away everyday items like clothing and shoes, jewelry, and children's toys and games is a great first step in keeping your pet safe. It is important to keep anything small enough for your kitten to swallow out of its reach, but even larger items can present problems.

Some kittens, particularly those that were weaned too quickly or suddenly, can develop a dangerous habit of sucking on cloth. The actual material can vary—from sweaters or socks to shoelaces. The kittens suck on the item, often to the point of ingesting a portion of the material. Once inside the digestive tract, the foreign substance can then cause life-threatening intestinal blockages.

By returning all items to their proper places, you also protect your family's possessions from being damaged by your pet. Like human infants, kittens go through a teething phase. Beginning around 11 weeks of age, your kitten will begin losing its 26 deciduous teeth. Over the next year, permanent teeth will replace these so-called milk teeth. Until the teething process is complete, however, your kitten's mouth will feel sore. Some kittens attempt to ease this pain by gnawing or chewing. You can help by providing your pet with appropriate chew toys—and keeping inappropriate items out of sight.

WE'RE CLOSED!

As you put your belongings away, make sure to close all doors and drawers securely. If you leave just one kitchen cabinet open, a curious kitten is sure to notice and venture inside it. Cupboards containing cleaning supplies or medications are of particular concern. Never assume that just because a bottle is childproof that a kitten cannot get into it. And always be careful when handling medicines or vitamins. Just one dropped pill could be lethal to your small pet.

It's not just closets and cabinets that kitten owners need to keep closed. Kittens love to climb into all sorts of

obscure spaces. Keep the doors to your washing machine and clothes dryer shut at all times, and always look inside these appliances to make sure they are kitten-free before running them. Also, make a point of keeping toilet lids closed to prevent your kitten from falling inside. A young kitten can drown in even shallow water. Use care with other basins such as sinks, tubs, and decorative fountains.

If your kitten will be an indoor pet, it is vital to keep exterior doors and windows closed. Even then, you must be vigilant. Kittens are crafty creatures, highly capable of learning how to undo latches. For this reason simply closing a screen door might not be enough to keep your pet safe. Yes, the latch is probably high above your kitten's head, but could your pet reach the handle by standing on a nearby piece of furniture or other object?

Screens, whether in windows or doors, are generally poor matches for a kitten's sharp claws. A screen may be enough to keep your kitten inside when you are there to supervise, but it is a wise idea to close windows and doors more securely whenever you can't watch your pet. If your kitten scratches a screen, be sure to replace it promptly. Otherwise it can provide an easy means of escape.

BEHIND THE SIGNS

Keep Out!

In some cases a closed door just isn't enough to keep a curious kitten out. If you find that your pet is among the many young felines that can open a cabinet, the next step is moving from kitten-proofing to child proofing. Kits including cabinet latches, electrical plug protectors, and corner cushions can be found in the baby sections of most department stores. All these items can be just as effective for kitten safety as for child safety. You probably won't have to use them indefinitely. In time your pet will find other, more acceptable ways to entertain itself. In the meantime, though, this simple step will help your kitten stay out of trouble.

NO STRINGS ATTACHED

Your kitten's playful nature will make almost any item a potential toy. Allowing your pet to bat a piece of crumpled paper across the floor is unlikely to cause any harm, but some objects should be considered off limits to your kitten at all times. Cords that operate window blinds or curtains hold a considerable amount of fascination for many young cats. But playing with window cords can be dangerous—many young cats have been strangled by these devices.

The best way to protect your kitten from window cords is to replace corded window treatments with cordless styles. Some windows even offer models that house blinds between two panes of glass with sliding levers for opening and closing. If replacing corded blinds isn't an option, tie the cords in knots and secure them as high up as possible. Moving furniture away from windows is a smart step, but this action alone isn't enough. If your kitten can reach the window sill, it can reach the cords—unless you place them out of reach.

Electrical cords are also a cause for concern with some kittens. The potential danger here is electrocution. To prevent your pet from chewing on these cords, treat them with a foul-tasting deterrent such as hot sauce. You can also find sprays for this purpose at your local pet supply store. Until you are certain that the problem is under control, however, be sure to unplug your electrical appliances when you aren't using them.

BEWARE OF BEGONIAS—AND *MANY* OTHER PLANTS

While it's true that cats rely mainly on meat for their nourishment, many kittens find plants simply irresistible. Your kitten may be most attracted to grass-like plants, or it may feast on any type of foliage.

Before you bring any type of plant into your home, you must be certain that it is safe to keep around your kitten. More than 700 different types of plants have been identified as toxic to animals. Even if your kitten doesn't appear to have an interest in plants, it is wise to keep any poisonous species as far away from it as possible. You simply never know when your pet might decide to take a bite.

Some types of plants may merely irritate your kitten, causing mild nausea or skin reactions. Other species might cause more intense symptoms, such as vomiting or diarrhea, if ingested. Some plants, however, are deadly. Because of your kitten's age and smaller size, it is even more vulnerable to toxic substances than adult cats.

Conduct a thorough inventory of your houseplants before your kitten's homecoming. Make a list of all the plants you own, and check to see if they are safe to keep around cats. A thorough list of species that are poisonous to cats can be found at the ASPCA website. Remember to include any outdoor plants on your property if your kitten will be spending time outside. If you will be providing your cat with an outdoor enclosure, be especially mindful of any vegetation within it.

If you would like to indulge your kitten's interest in plants, consider buying one or more types of cat grasses. Catnip is just one of the many types of grass that is edible for cats with a fondness for foliage. Catmint, parsley, and wheatgrass are also safe for feline consumption. Having cat grass available for your kitten may also deter your pet from chomping on your other houseplants. Even though your parlor palm is entirely safe to keep around your kitten, you still probably don't want your pet making a meal out of it.

KITTEN-FRIENDLY SUPPLIES

FROM PICKING A CRATE TO MUST-HAVE TOYS

hether you buy a kitten from a breeder or adopt one from a shelter, your new pet will need a few basic items right away. Shopping for these necessities before you bring your kitten home will help make the transition go as smoothly as possible. It will also free you up to spend more time with your pet instead of running errands during those first few days following your kitten's homecoming.

GOING ALONG FOR THE RIDE

One item that you must pick up before taking your kitten home is a crate. You will also need to use it again as soon as you head to your kitten's first veterinary appointment. Consider this carrier like a car seat for your kitten. It will help keep your pet safe while riding in your vehicle. Kittens that will be shown in confirmation or taken to a professional groomer regularly will use their crates frequently, but this item ranks at the top of the essentials list even if your kitten will be spending most of its time at home. In the event of an emergency such as a fire or a natural disaster, this single item will allow you to keep your pet with you as you move to a safer area. Some owners simply use a collapsible cardboard carrier to take their cats to the vet once a year, but most people prefer a sturdier crate for a couple of reasons. First, your kitten may not enjoy riding in an automobile as much as you hope. With a little luck, your pet will adjust to this new experience easily and quickly, but the fact is that some cats simply hate riding. If your kitten falls among the latter group, it may try to force its way out of the carrier. A tiny kitten is unlikely to succeed at this task, but as your pet grows, escaping a glorified cardboard box becomes a greater possibility. The nondisposable types of carriers also make it possible for your kitten to view its surroundings— which just might make your pet enjoy the ride a bit more.

When it comes to crates, you will have many choices. Which one is best for your kitten depends largely on how often you will be using this item and where you will be taking your pet. If you plan to travel with your kitten, for example, you will need a rigid carrier that is deemed suitable by most airlines. When flying with your pet, you will be able to keep it in the plane's cabin with you, but its crate must be small enough to fit under the seat in front of you, so size is also a factor for frequent travelers.

Kitten Fact

A smaller carrier is more likely to make your kitten feel at ease while traveling, be it on a quick trip to the vet or a longer journey.

Make sure the crate you choose has enough room for a set of dishes so your kitten can eat and drink comfortably, but you don't want the interior to be too spacious. A general rule is that a carrier should be no more than twice the size of the animal—just big enough for your pet to stretch, turn around, and lie down while inside it. Since your kitten will be doing a lot of growing over the next year or two, though, you may want to select its crate based on its expected adult size.

Crates can range significantly both in style and price. If you want your kitten to have a designer carrier, you are sure to find a model that looks more like luggage than a crate. Rest assured that a less pricey model will serve your pet just as well, however, and possibly even better. A hard plastic carrier is much more difficult to escape than a soft-sided one. The best way to make your kitten's carrier more comfortable for your pet is to add a soft blanket or crate liner.

BEHIND THE SIGNS

Not Just for Travel

If you want your kitten to be at ease inside its crate, make this item part of its everyday scenery. Taking the carrier out only when it's time to visit the veterinarian can easily lead to anxious feelings whenever your kitten sees the enclosure. By leaving the carrier out with the door open, however, you can turn a potentially frightening item into a fun hideout, or even a sanctuary. To encourage your kitten to venture inside the crate, place a treat or two inside it, but be patient. Let your pet explore this item on its own timetable, and you just might find that it becomes a favorite hangout.

EAT, DRINK, AND UM . . .

Before taking your kitten home, ask the breeder or shelter what kind of food your new pet has been eating. You may choose to change to a new formula or keep feeding the same brand, but even if you plan to swap your pet over to a new food, it is wise to pick up a small amount of your kitten's current food. Making sudden dietary changes may upset your pet's stomach, causing diarrhea. For this reason introduce a new food slowly, mixing it in with the old brand and gradually increasing the ratio.

Your kitten will also need dishes for both food and water. When it comes to these items, it is often best to keep things simple. Do make sure that the bowls you select are wider than the span of your pet's whiskers. From the time they are kittens, most cats prefer that their whiskers not touch the sides of their dishes.

Stainless steel dishes make an ideal choice. They are reasonably priced, easy to clean, and won't break if you accidentally drop them. Fancy ceramic dishes may be pretty, but if they hit a tile floor, there will be no putting them back together. Ceramic can also contain lead. Pet dishes aren't held to the same safety standards as those made for people. If you must buy ceramic items, choose bowls made for human use for this reason.

Plastic dishes, while inexpensive, are also problematic. Plastic can cause allergic reactions in many kittens. If your kitten appears to have acne on its chin, it may have been caused by eating from a plastic dish. Plastic also scratches easily, and those tiny grooves are just large enough to harbor a surprising amount of bacteria, even after you wash the dishes.

About 20 minutes after your kitten eats, it will need to use its litter box, making this another important item to have waiting for your new arrival. A simple litter box works fine for most kittens, although some prefer models with hoods for privacy. If you opt for a hooded design, though, it might be wise to go with a removable hood, as a fair number of kittens are also intimidated by this feature. If your kitten has already begun housetraining, you may want to go with the type it has become accustomed to using.

Kitten Fact

Buying two sets of dishes will help ensure that your pet always has a clean set waiting when dirty dishes head into the dishwasher.

Beware of Yarn

Despite the cliché image of a kitten playing with a ball of yarn, extreme caution must be used when allowing your pet to play with this material. Even if you supervise your kitten when it plays with yarn, it can inadvertently swallow the material before you can intervene. And unfortunately, many kittens respond to this problem by swallowing even more. Whether long or short, any length of yarn (or string) that your kitten swallows can cause dangerous blockages in its intestinal tract. It can even wrap around a section of the intestines, cutting off vital blood supply.

Automatic litter boxes make cleaning easier for owners, but they are considerably more expensive. Also, remember that the motor will have a life, making it necessary to replace an automatic model at some point in the future. A conventional litter box, on the other hand, should last your kitten's lifetime. Finally, don't forget to pick up some litter. You will need to keep about 2 inches (5 centimeters) of this material in the box at all times.

HAVING A BALL

It's no secret that most kittens love to play, but much like children, playtime for young cats is always more fun when they have a playmate. Toys like small balls, laser pointers, and fishing poles with dangling feathers top the list of interactive toys that you can use with your new pet.

Because they encourage your kitten to run and jump, toys like these can help provide your kitten with a great deal of exercise, which is crucial to its good health.

Toys can also be useful for entertaining your kitten when you can't be with your pet. A ball that lights up whenever your kitten moves it can encourage it to chase this toy around the house even without someone there to roll it. Likewise, a battery-powered toy mouse can provide a thrilling chase for your tiny hunter.

You needn't invest a lot of money to find toys your cat will enjoy. Sometimes you won't have to spend any at all. An empty paper bag, for instance, can hold a curious kitten's attention for hours. Your pet may play with this item alone, or you may heighten the fun by tapping on the outside, tempting your kitten to bat at this unknown object.

A BED OF ITS OWN

Whether you want your kitten to sleep in your bed or not is a matter of personal preference. Even if your kitten has also claimed a particular pillow, though, it may be wise to provide your pet with a bed of its own as well. A soft, comfortable bed can make an ideal spot for your kitten to take naps throughout the day. When it comes to this item, look for two major selling points. First, it must be soft. Most kittens love the feel of fleece or velvety material on their paws. Second, it should be machine washable. Your kitten, of course, will care much less about this.

WELCOMING KITTY HOME

INTRODUCING YOUR HOME, CHILD, AND OTHER PETS

our kitten's homecoming day is bound to be an exciting one for both of you. Perhaps you have been waiting for your new pet to be old enough to leave its mother, or maybe the animal shelter needed a few days to review your application before approving it. Either way, to you it probably feels like time has been dragging. But remember that your kitten hasn't been able to prepare for the adventure that awaits you both. Your new pet won't even realize that it is going home with you until the two of you leave together.

Your kitten is going to love you, and you are going to have a great deal of fun together. The first day—and night—in your home, though, may be a little scary for your young cat. You can help reduce your pet's anxiety by taking a few simple precautions. Having your plan in place before the big day arrives will also make things less stressful for you and everyone else in the household.

FIRST THINGS FIRST

If you have purchased a bed for your kitten, decide where it will go before bringing your pet home. Select a quiet spot away from the household's heaviest foot traffic. It's hard to rest when people are whizzing by constantly. Also, make sure the location you choose doesn't expose your kitten to any drafts. Young kittens must be kept warm. Even in the middle of summer, a young cat can feel chilled—by window breezes or air conditioning, for example.

You should also have your kitten's dishes and litter box ready before your pet's arrival. Many owners place a kitten's bed in the bedroom, its feeding dishes in the kitchen, and the litter box in the bathroom, but during the first few weeks, it is best to keep everything your kitten needs in the same general area. Doing so will make your pet feel more secure and also make housetraining easier.

Kitten Fact

To ensure the good health of animals sharing a household, many veterinarians recommend instituting a quarantine period after bringing a new kitten home. The standard length for this isolation period is 14 days— just long enough to make sure that no serious illnesses are passed between the resident pets and the new arrival.

THE MEET AND GREET

Although everyone in your household may
want to welcome your new arrival,
meeting the whole family at once may be
too much for your kitten. Instead, focus
on getting your pet settled before
allowing people to approach it. If you plan

to limit your kitten to a single room
initially, go there promptly. After closing
the door, allow your kitten to explore
the room as much as it likes. Your pet
may roam from one area to another,
investigating everything it sees
thoroughly, or it might simply curl up

and take a nap. The latter possibility is more likely if the ride home was a long one. If this is the case, be patient while your kitten rests; young cats need lots of sleep.

Slowly, allow one family member at a time into the room to meet your kitten. Keep the introductions short—a simple hello and gentle pat on the head is ideal. Encourage each person to allow the kitten to come to him or her on its own. Children in particular may scare a kitten by moving too quickly or making too much noise. And a bad first impression could have a negative effect on their future relationship with the animal. Explain this to your kids before homecoming day, but a gentle reminder may also be a smart idea.

Older children can be taught the proper way to treat a kitten, but toddlers and preschoolers are simply too young to understand how easily they can hurt a kitten, albeit unintentionally. For this reason your kitten's interactions with young children must be limited. Allowing a young child to pat the kitten while you hold it is fine, but letting the child hold the animal is simply too risky.

Give school-age kids a quick lesson on the proper way to hold a kitten. Demonstrate by placing one of your own hands behind the animal's front legs and the other under its hindquarters. Explain

that they must always be gentle when interacting with kittens, never pulling their ears or tails. You may trust a well-behaved preteen alone with your kitten, but younger kids should be supervised until you are certain they understand how to be respectful of the small animal.

Never leave your kitten alone with anyone, regardless of age, if you suspect the person won't treat it properly. You should feel comfortable with everyone you allow to interact with your pet—from your children's friends to groomers. Listen to your intuition, and watch your pet around new people. Your kitten may be young, but animals can be better judges of people than many humans. If your pet doesn't like a particular person, it might be worth taking a closer look as to why.

Kitten Fact

A kitten will move away from something or someone it fears. In the most overt cases, it may even twitch or tense up its muscles. Pay attention to these signs and give your pet some space to help it feel more at ease.

FOUR-LEGGED INTRODUCTIONS

If you have other pets, you will need to introduce them to your new kitten as well. The exact procedure will depend on the species and personalities of the pets involved, but each introduction should be carried out slowly. Once your kitten has settled into your home a bit, it can then meet these other animals, one at a time.

A large dog—even if it is the friendliest on earth—can easily frighten a young cat. A dog could also seriously injure your pet by biting or jumping on it. For this reason, attach your dog's leash before the two of you enter your kitten's room. You may also want to place your kitten inside its carrier for the introduction. Doing so will prevent the kitten from running away, an action that could trigger a chase.

If you already have another cat, follow the same basic procedure above, minus the leash. The older cat may show little interest in the new kitten, or it may be obsessed with sniffing the new addition from head to toe. As long as both cats act friendly, continue to allow them to get to know each other a little more.

Keep initial interactions brief, and praise both parties for good behavior. As your pets become more comfortable, you can allow more contact between them, but remain vigilant until you are confident that the animals are tolerating each other well. At the first sign of aggression, remove the resident pet from the room. Don't despair if this happens. It simply means that you must try again later, ideally in another day or two when everyone is calmer.

EXPANDING THE CIRCLE

When it comes to introducing your kitten to nonhousehold members, be patient yet persistent. Once your pet has become comfortable with you and the other people in your home, start introducing new people. Inviting family and friends to visit your home regularly is a great way to accomplish this task. During the first few weeks, limit your company to people who truly love cats, for they will provide the highest potential for positive experiences. If your kitten appears to be frightened, consider whether an arbitrary object is the cause. Perhaps your pet has simply never seen a man with a beard, a person who wears eyeglasses, or someone who uses a wheelchair. Hats can also intimidate a kitten at first.

Although it can be tempting to keep people away from your pet if it shows fear around them, postponing socialization until your kitten is older may just make it more difficult. Overexposure isn't the answer either. Forcing interactions or bringing too many people into your home at once will overwhelm your pet. Instead, keep exposing your kitten to the people in your life, gradually. It may take several visits from the same person before your kitten warms up.

MOVING AROUND THE HOUSE

Once your kitten has had some time to adjust to being in your home, you can introduce your pet to the rest of its new home. This process too should be undertaken slowly. After placing your kitten in its carrier, take your pet to another room. After closing the door, place the crate in the center of the new room and open the door so your kitten can explore. If your pet chooses to stay inside the carrier, let it be. More likely, though, your kitten will want to see all the new things in this new space.

Remain in the room with your pet throughout this investigative process. Kittens can easily slip underneath furniture or behind appliances like refrigerators remarkably quickly. Each room should be properly kitten-proofed before you allow your new pet to go exploring, but supervising can help you avoid an extensive search when it's time to return your pet to its own area.

The number of rooms you cover in a day will depend on your pet's energy and comfort levels. Watch your kitten for signs that it is tired, and call it a day when you think your pet has had enough. By gradually introducing your kitten to your home and all its inhabitants, your new pet will settle into your household and become as much a part of it as everyone else.

CALMING A NERVOUS KITTEN

If your kitten has just left its mother and siblings for the first time, it may have a hard time eating and sleeping at first. The following tips can make the first few days and nights a little easier on your pet:

You can entice your kitten to eat by warming up its food. Heating food releases the scent, making it more appealing. Be careful not to overheat it, though—test the temperature with your finger before offering it to your pet.

Your kitten will have slept curled up with its siblings. To make sure your pet is warm enough in its bed, invest in a stuffed animal that can be heated in the microwave. It shouldn't be hot, but a warm, soft friend may provide comfort in more ways than one.

Wrap a small ticking clock in a towel, and place it in your kitten's bed. The sound will mimic its mother's heartbeat—and hopefully lull your pet to sleep.

No. 8 A MEAL FIT FOR A KITTEN

KNOW WHAT AND WHEN TO FEED

At one time cat owners simply picked up a bag or a few cans of a nondescript cat food each time they visited the grocery store. Over the last several decades, however, many owners have learned that the health of their feline companions greatly improved when the owners became better acquainted with and more selective about the ingredients in their pets' food. The proper ratio of certain nutrients delivered through quality ingredients can make a huge difference in how a cat looks, feels, and acts.

A cat's nutritional needs vary from those of other companion animals. Cats, for example, need more meat in their diets than dogs. A kitten's dietary requirements are even more specialized than an adult cat's, making the selection of your new pet's food formula an important undertaking.

H₂OH!

When most people think of feline nutrition, they start by considering how much protein, fat, and carbohydrates that a particular food formula offers. Vitamins and minerals often top the list of owners' concerns as well. But did you know that your kitten's most important nutrient contains no vitamins or calories whatsoever? By leaps and bounds, water is the most crucial component to a healthy feline diet.

Water does numerous things within the feline body. First and foremost, it prevents dehydration, a dangerous condition for any animal. Water also transports other nutrients throughout your kitten's body. Next, it helps remove wastes in the form of urine. And finally, it helps regulate your kitten's body temperature.

The bad news is that kittens are far from sponges when it comes to drinking water. A cat's thirst drive simply isn't as strong as that of other animals. For this reason it is essential that you make fresh water available to your new pet at all times, ideally in several places throughout your home. Just seeing a bowl of water can serve as a gentle reminder for your kitten to take a drink.

Since leading a kitten to water isn't always enough to make it drink, however, you can increase your pet's water intake by feeding it wet food. Most canned kitten foods are at least 75 percent water. Find a variety your kitten likes, and you will virtually ensure that your pet stays properly hydrated.

One thing you shouldn't give your kitten to drink is milk. Despite the common belief that milk is good for cats, many are actually lactose intolerant. Kittens that drink milk can suffer from a variety of stomach issues, including diarrhea and vomiting. Plus, your kitten gets the same nutrients found in milk from its own food, so there is no benefit to offering it even if your pet tolerates it well.

BREAKING IT DOWN

The biggest difference between a kitten's dietary needs and those of an adult cat is the higher number of calories a kitten requires. Kittens can easily double or even triple their weight during the first few weeks of their lives. All this growing is accompanied by an intense energy level, making it nearly impossible for the animal to get enough calories out of just one or two meals a day. Cats in general enjoy eating throughout the day, but this habit is especially prevalent in younger members of the species.

The list of nutrients that your kitten's body needs isn't much different than what adult cats require. The ratio of these nutrients, on the other hand, must be adjusted for younger cats. Kittens need more protein than older felines. Exactly how much more is a subject that is often debated among feline nutritionists. A recent study by the University of Illinois published in the *British Journal of Nutrition* found that kittens placed on high-protein diets had lower levels of beneficial bacteria such as *Bifidobacterium* and *Lactobacillus* in their lower digestive tract than those eating formulas with more balanced protein ratios.

A reasonable conclusion may be that while kittens need more protein than their elder counterparts, they still need other nutrients such as fat and carbohydrates as well. Many owners believe the widely held opinion that cats do not need carbs at all, but a large number of veterinarians actually recommend carbohydrates for kittens. Carb critics point out that wild cats do not consume this nutrient; their intestines aren't even equipped to digest them. Domestic species, however, have longer digestive tracts than wild cats. They can obtain a fair amount of energy from cooked complex carbohydrates, such as whole grains.

The key is choosing the right type of carbs. If your kitten can't digest a carbohydrate easily, the benefits will be lost. Furthermore, poorly digested carbs can have unpleasant side effects. Kittens that consume too much lactose, for instance, may suffer from gas, bloating, and diarrhea. Simple sugars are also bad for your cats.

When selecting a food for your pet, look for a high-quality brand and a formula made specifically for kittens. Some foods may claim that they provide adequate nutrition for cats of all ages, but don't be fooled by this impossible assertion. Your kitten should remain on a kitten formula until it is about a year old.

You are unlikely to find a nutritious kitten food on the shelves of your local grocery store, but it is important to realize that not every brand sold by pet supply stores is

☙ Kitten Fact

Occasional treats won't hurt your kitten, providing you offer them in moderation. Aim at keeping the calorie count for treats below 10 percent of your pet's total daily intake. Also, remember that treats can be healthy. The company that makes your pet's food probably makes some tasty treats, too.

ideal either. Neither a high price nor a popular name guarantees quality. The only way to ensure that the food is a healthy one is by reading its label.

Many pet food companies use filler ingredients in their products for the sake of boosting their profit margins. Nutritious proteins like chicken and turkey can be expensive, but byproducts are cheap. These low-grade ingredients such as beaks and feet aren't considered fit for human consumption, but they are allowed to be used in pet foods. And because byproducts can technically be classified as protein, including them in the ingredients lists makes a food seem healthier.

Wet or Dry Food?

Feeding dry food exclusively can lead to bladder and urinary tract problems in adult cats. For this reason it is wise to steer clear of a dry-only diet for your kitten as well. If you do feed your young pet dry food, be sure to supplement it with regular canned food. But you may have an even more important reason to skip the dry food altogether—at least until your kitten is several months old. A kitten's tiny baby teeth may not be large or strong enough to chew dry food. Offer your kitten wet food until its permanent teeth come in to ensure that it receives proper nutrition in the meantime.

A closer look and a better understanding of how some companies push the nutritional envelope can make it easier to separate the quality foods from the imposters. Look for foods with no byproducts, no bone meal, and no chemical preservatives. Like byproducts, bone meal is considered a source of protein, but it is far inferior to lean meats. If you opt to feed your kitten canned food, it probably won't contain preservatives. Dry food, on the other hand, needs some type of additive to give it a reasonable shelf life. The word preservative has a bad reputation, but vitamin-based preservatives called tocopherols are considered much safer than their chemical alternatives. If a food is preserved with Butylated Hydroxyanisole (BHA), Butylated Hydroxytoluene (BHT), or ethoxyquin, keep looking for a healthier option.

IT'S ABOUT TIME

Once you have picked out your kitten's food, the next step will be setting up a feeding schedule. Feed your kitten three to four times a day until it is six months old. To help your pet get the most out of its meals, space them as evenly as possible.

At six months you can transition your kitten to two daily feedings. This routine will carry your pet through the remainder of its first year.

When you take your kitten for its first checkup, ask the vet how quickly it should gain weight over the next few months. You needn't weigh your pet every day, but do keep an eye on the scale to make sure that your kitten is getting all the nutrients from its food that it should. Refer to the package label to determine exactly how much your pet should be eating each day.

If your kitten gains too much weight, you may need to cut back on the amount a bit. If it isn't gaining enough, however, schedule another appointment with the vet. This is your best ally in creating a plan to fix the problem. Free feeding can lead some animals down the road to obesity, but for an underweight kitten, providing access to food at all times can be a lifesaving solution.

THE BIGGEST MISTAKES

Avoid these common feeding errors to keep your kitten as healthy as possible:

OVERFEEDING

This common error can lead to obesity, which can simultaneously cause a variety of serious health problems. Watch your kitten's portions and adjust them accordingly if your pet starts gaining too much weight.

OVERDOSING ON VITAMIN SUPPLEMENTS

If your kitten is eating a balanced diet, vitamin supplements shouldn't be necessary to maintain your pet's good health. If you still want to add vitamins to your kitten's feeding regimen, check with your veterinarian first to make sure you know the right type and dose for your pet's age and size.

FEEDING MEAT AND FISH ONLY

Kittens love meat and fish, but if they don't get other foods as well, they can develop nutritional deficiencies. Home-cooked diets for kittens can be especially tricky if calcium and phosphorus aren't provided in the right balance. When not planned properly, these diets can lead to a condition called nutritional secondary hyperparathyroidism. Likewise, too much fish can lead to an illness called steatitis, also known as yellow-fat disease.

$\mathfrak{N}o.$ 9 **GROOMING KITTY**

BRUSHING, BATHING, AND BEYOND

$\mathfrak{M}$ost kittens require only minimal grooming during their first few weeks or months to keep them clean and healthy. The bigger purpose of grooming young cats, especially those that will need extensive grooming as an adult, is to get them used to being brushed, bathed, and otherwise coiffed. Kittens exposed to regular grooming tasks are usually far more tolerant of grooming as they move into adulthood.

FACE THE FEAR

The biggest hindrance to most grooming tasks is fear. You may have taken great pains to pick out your kitten's brush—the one with the slanted head and bristles that massage your pet as you brush. Still, your kitten may see this unfamiliar object coming at it and instantly assume that a brush is some sort of torture device that it must evade immediately. The trick is introducing the brush and other grooming equipment in a way that makes it as least threatening to your pet as possible. Your kitten's first grooming experience will set an important precedent for the ones that follow. (No pressure, right?)

You also might have to face some of your own fears while grooming your kitten. Perhaps you worry about injuring your pet when trimming its claws. Or maybe you are afraid that your kitten will scratch you during a bath. You might even worry about your kitten viewing you as the villain for being the person in charge of grooming. Whatever your concerns, the best way to alleviate them is educating yourself before beginning any grooming task. Trying to read this book while holding your kitten in one hand and nail scissors in the other, for example, is a recipe for disaster.

BRUSHING UP ON BRUSHING (AND COMBING)

The best way to take the grooming pressure off yourself is removing it from your kitten as well. Since the pervading goal of grooming a kitten is creating a positive experience, you have the luxury of being able to focus more on your pet's comfort level than on the technical results. Whether you are brushing your kitten or performing another grooming task, begin by allowing your pet to investigate the tools.

Set the brush down in front of your kitten. Encourage your pet to check out the brush, but don't force it to touch it. Just having your kitten in the general vicinity of the grooming tool without showing signs of anxiety is an excellent start. Many kittens will sniff at the new object and then move on to something else that piques their interest.

Do be careful with sharp or otherwise dangerous items, however, such as scissors or a mat breaker. You may find the latter tool helpful if your kitten is a longhaired breed with a profuse coat, like the Turkish Angora. Shorthaired kittens won't need this advanced grooming tool, but any object that can harm your pet should be kept out of its reach for reasons of safety.

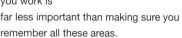

Once your kitten has accepted the presence of the brush, try picking it up and moving it gently down your pet's back. Pay attention to body language to judge your kitten's reaction. If your pet stretches out and relaxes during brushing, take this as a sign that your pet is enjoying itself. Many kittens delight in being brushed, but in the beginning it is important to keep grooming sessions short, ideally just a minute or two. You can gradually increase this length over time.

As your kitten becomes more relaxed with being brushed, move on to other areas. Breeds with more profuse coats will be prone to the formation of mats on their bellies and legs, particularly where the legs meet the body. You can keep them from morphing into more serious snarls, however, by performing thorough daily brushings. A complete brushing should include the back, belly, legs, ears, and tail. The order in which you work is far less important than making sure you remember all these areas.

Brushing removes dead and loose hair as well as dirt from your kitten's coat. In order to accomplish this part of the task, however, you must reach the skin. For shorthaired cats, a slicker brush alone is usually sufficient, but for breeds with longer or thicker coats, it is smart to follow up brushing with combing. Using a comb will help identify any areas that you may have missed during the brushing process. If you do find a snarl, be gentle in untangling it. Pulling your kitten's fur can make your pet fearful of future grooming sessions.

If your kitten doesn't respond well to a conventional brush, consider buying a glove-style brush. As the name implies, this grooming tool is worn on the groomer's hand. Lined with rows of soft rubber nubs, the glove brushes fur without the risk of scratching the kitten. A brushing with a grooming glove can feel a lot like being massaged by hand, partly because that is essentially what's happening.

⊛ Kitten Fact

To keep your kitten's fur from developing mats, always brush its coat before bathing. Even tiny knots can be almost impossible to remove once they become wet.

A TOUCHY SUBJECT

Even when you aren't actually brushing your kitten, you can work on teaching your pet to tolerate this important grooming task simply by touching your pet. Many kittens that resist brushing do so because they aren't accustomed to having specific parts of their bodies handled. Ears, feet, and tails often top the list of areas that can trigger negative reactions when brushed. To avoid this problem, make a point of gently massaging your kitten from head to tail during nongrooming times. If you regularly rub your kitten's paws, your pet will quickly learn that there is nothing to fear from your brushing them either. It may even grow to enjoy both activities.

RUB-A-DUB-DUB

Cats, even young ones, are skilled at keeping themselves clean. Unless you plan to show your kitten, regular bathing won't be necessary for most breeds. Bathing a kitten too often can even harm your pet's coat, stripping it of important oils that keep it feeling soft and looking beautiful. Certainly, if your kitten gets into something messy, a bath may be necessary, but the best plan for bathing is doing it as infrequently as possible.

To bathe your kitten, gather all your supplies before involving your pet. The worst time to realize that you have forgotten the shampoo is after you have placed your kitten in its bathwater. Fill the kitchen or bathroom sink with about 4 inches (10 centimeters) of lukewarm water, and have your kitten's shampoo, a plastic cup or sprayer, and an absorbent towel nearby before retrieving your pet. It is also smart to adjust the thermostat to make sure your kitten won't feel chilled when the bath is finished.

Patience is fundamental to helping your pet adjust to grooming tasks like brushing, but for bathing, it's all about getting the job done as efficiently as possible. As soon as you place the kitten in the water, gently wet its fur using your cup or sink sprayer. Be careful not to get any water in its eyes. Use a damp cloth to wipe its face. Next, place a small amount of shampoo in the palm of one hand, then rub both your hands together to form a weak lather. Only use shampoo formulated especially for kittens; your own shampoo is much too harsh for your kitten's sensitive skin.

Work the shampoo into your kitten's coat using your fingers, and leave it on for the time specified on the product's label. Next, rinse your kitten thoroughly. If you leave any of the soap behind, it could aggravate your pet's skin, so rinse twice to make sure you have gotten every trace of the shampoo out. Dry your kitten gently with a soft towel. Longhaired breeds should be brushed immediately after a bath to keep snarls from forming.

🐾 Kitten Fact

While many cats detest the water, some breeds enjoy it. The Turkish Van is known for playing in water. Even other breeds and mixed breeds can be taught to enjoy the water if they are exposed to it at a young age. A cat that is introduced to water as a kitten may not fear it as an adult.

TIME FOR TRIMMING

Whether your kitten is purebred or a mixed breed with long hair or short, it will need its nails trimmed regularly. Keeping your pet's claws clipped offers several advantages. The biggest of these by far are your pet's comfort and your furniture's longevity. Kittens with overgrown nails tend to damage items like curtains and upholstery far more than animals that receive routine pedicures.

To make trimming a bit easier on both you and your kitten, enlist some help for this grooming task. Have another household member hold your kitten while you do the clipping—that way you can focus on what you are doing instead of trying to keep your pet still. This is another task in which early exposure will pay off in the long run. When a young kitten learns that there is nothing to fear from nail trimming, that same animal will tolerate trims much better as an adult.

Begin by gently pressing your kitten's foot between your thumb and index finger, just enough to extend the nails on that foot. Next, using scissors made specifically for this task, snip away just the end of the first claw at the spot where the nail starts to curl. Repeat this procedure for all the other nails, but be respectful of your kitten's comfort level. It is better to clip the nails on just one foot and end on a positive note than to force your young pet to have all its claws clipped at once. Spreading the task over several days can even be a practical way to get your kitten used to regular nail trims, since this task will only be necessary every few weeks.

Avoid the quick, the pinkish region of the nail that includes blood vessels. If you cut into this area, it will bleed. It will also be painful, which can definitely leave your kitten feeling fearful of future trims. If you do snip the quick, apply styptic powder to the wound at once to speed clotting. If you repeatedly make this mistake, it may be better to leave nail trimming to a professional groomer or veterinarian.

No. 10 KITTEN SOCIALIZATION

UNDERSTANDING WHY AND HOW

Many different things will influence the temperament of your kitten. Genetics will undoubtedly play a role, as most kittens inherit various personality traits from both their parents. Certainly, each kitten is also an individual with qualities of its own that will emerge over time, regardless of its parents' mannerisms. Spend a little time with any litter of kittens, and you will see that each one is at least a bit different, even though they all have the same dam and sire.

NURTURE AND NATURE

Just as powerful as genetics, however, is a kitten's environment. How a kitten is treated during the first few weeks of its life specifically can have an overwhelming effect on how the animal interacts with people, fellow cats, and other animals throughout its lifetime. A 2008 study by Dr. Rachel Casey of the University of Bristol revealed that properly socialized kittens make better pets, forming stronger bonds with their owners. Owners of socialized kittens reported "significantly higher emotional support from their cat approximately 10 months after homing."

 Kitten Fact

While you want to keep your interactions with your new kitten positive, you mustn't allow your pet to behave badly. Letting your kitten bite or scratch you—even playfully—will only teach it that these behaviors are acceptable. If your kitten touches you with its teeth or claws, say the word no, and move in a way that prevents your pet from continuing the behavior. If necessary, distract it with a favorite toy.

FIRST IMPRESSIONS

One of the most crucial periods of socialization for your kitten will have already taken place by the time your new pet joins your family. Between two and seven weeks of age, kittens absorb a great deal of information from their environments. Their brains are still developing, and they use their early experiences as a model for the future. Kittens that are handled respectfully by people during this time period learn that there is nothing to fear from humans.

However, kittens that are mistreated during this time period come to expect hurtful behavior from people. Since they have no other experience with humans, they assume that all people must be abusive. As a result, they will withdraw or become defensive as a means of protection in the future, no matter how lovingly a person may treat them.

Kittens that are rescued shortly after birth—or born into a rescue situation—can have a harder time bonding with people if rescue workers do not make proper socialization a priority. This is why foster care is an ideal scenario for

homeless mothers and their litters. If you volunteer in this capacity, you must find that all-important happy medium when it comes to the approach and handling of the kittens, though, in order for the experience to have the desired effect.

For newborn kittens socialization begins by winning over the trust of the mother cat. You needn't bribe her with tasty treats—although that may not hurt. All you truly need to do is show her that you mean no harm. Without her permission, socializing the kittens will be a much more difficult task. Mother cats can be wildly overprotective, and if the mother cat has had negative experiences with humans herself, that sad fact will likely only complicate matters.

 Kitten Fact

Don't despair if your kitten didn't receive proper socialization before you entered its life. Even older kittens that missed out on early socialization can usually be socialized to people—it just might take a little extra time and work.

Just What the Doctor Ordered

Handling a kitten helps socialize it to people. This task can also make routine veterinary care a whole lot easier. As you handle your young cat, examine the animal as a vet might do during the initial part of a checkup. While holding your kitten, gently turn it over, inspecting both its back and belly. Lightly run your fingers over the animal, touching its ears, feet, and tail. By making this kind of handling a familiar event in a kitten's life, you will help your pet become more receptive to being touched in the same ways by the veterinarian. Don't just fake an exam: it is the perfect opportunity for a health check.

When the kittens are two weeks of age, socialization may simply consist of stroking them regularly and picking them up only briefly. Returning the kittens to their mother promptly at this stage helps strengthen the trust you have begun building with her. It is essential to repeat this exercise many times each day, however. You can gradually increase the amount of time you handle the kittens. The most important thing is to make the experience a positive one.

Ideally, at least four different people should take part in this ongoing exercise. And they should be as different as possible in terms of age, gender, and appearance. Kittens handled by men, women, and children at this stage will be more comfortable in the company of a wider group of people than those that only have experience with women, for example. Remember, the people and things a kitten encounters at this age will become part of its frame of reference for what is normal.

WE'VE ONLY JUST BEGUN

While a kitten's earliest weeks have been identified as the all-important beginning of socialization, exposing your kitten to new people and things must be an ongoing process if you wish for your pet to grow into a well-adjusted cat. It is essential to understand that the positive effects of early socialization can be reversed if owners do not make this task a priority. The more effort you put in, the friendlier your kitten is likely to become.

When we hear the word socialization, we instantly equate it with people and other beings such as fellow cats and dogs. Interactions with people and other pets should certainly remain part of the process, but once you have introduced your kitten to its housemates, you can also start expanding your pet's exposure to a variety of new things that will also make it more comfortable with the world around it. Some of these things will be literal—objects that your kitten may encounter and should not fear. Other things may be intangible yet still prevalent in its environment, such as noises.

If kittens are only exposed to a limited number of objects, or type of objects, they can become surprisingly set in their ways. When you enlarge your kitten's radius in your home from that initial room where it spent its first few days, for instance, consider giving it access to a room with different sights, sounds, and textures. If at first you kept your kitten in your carpeted bedroom, make the next room one with tile or hardwood floors. Various textures will feel different under your pet's feet. If you ultimately plan to place your kitten's food dishes in the kitchen, your pet must be comfortable walking on the floor in that room.

You can expose your kitten to a variety of sounds by turning on televisions, radios, and computers in its presence. It is also important to expose your kitten to the sounds of mundane appliances such as washing machines and vacuum cleaners. Let's face it, homes can be noisy places. From kitchen mixers to hairdryers, daily life involves a whole lot of noisemaking. Maintaining your normal routines will help your kitten adjust to your home instead of vice versa. Waiting to run the dishwasher until the end of your kitten's nap might seem like a conscientious gesture, but doing so could cause your kitten to need total quiet when resting in the future.

BEHIND THE SIGNS

Hey, That Smells Familiar...

You can make use of your pet's highly developed nose by bringing new scents to your kitten, when your pet can't go to the sources. Maybe your child must leave for summer camp just before your kitten comes home. If you want to make the meeting a less stressful one for your pet, have your son or daughter sleep with a blanket that your kitten can use after it arrives. This technique can also be helpful if you plan to add a dog to your household in the future. You might consider rubbing a clean cloth on the coat of a neighbor's dog and then placing it in your kitten's vicinity. Be sure the other animal is healthy and has been vaccinated.

To avoid getting locked into buying only certain types of food, toys, and cat litter, expose your pet to a variety of products for a while. This approach not only helps create a more flexible pet, but it also gives you a chance to evaluate various brands and formulas firsthand before settling on which ones you think are the best. No matter how much your kitten enjoys eating food made from tuna, for example, eating just this variety can lead to an outright refusal to eat anything else down the road. Tuna in particular has such a strong smell and taste, it can actually be addictive for some kittens if it is not rotated with other types of food.

Finally, as you work on exposing your kitten to all these new things, don't forget to keep socializing your pet to new people. A kitten that sees new faces and hears new voices on a regular basis will be more self-confident and open to new experiences as it moves into adulthood. It may even be more likely to bond with a new member of the family such as a new spouse or another pet.

KITTEN KINDERGARTEN?

Many new puppy owners waste no time in signing up for classes to help their pets become properly socialized, but no such classes have existed for new kittens—until recently, that is. An Australian veterinarian named Kersti Seksel has started a program called Kitten Kindy™ that offers a series of classes in which kittens can learn to interact with people and fellow felines. The idea has been so well received that the American Association of Feline Practitioners' Behavior Guidelines now includes information about taking classes like this one. Ask your local humane society if one is offered in your area. For safety's sake kittens taking part in a class must be vaccinated, dewormed, and FeLV/FIV-tested to take part. In addition to the socialization a class offers, it also provides you with an opportunity to show your pet that car rides don't always lead to a vet appointment.

KITTEN DEVELOPMENT

FROM ONE WEEK TO ONE YEAR

*O*nce your kitten has grown into an adult cat, you will probably wonder where the time went. To many owners kittenhood seems to fly by, but each stage is different and important in its own way, making it essential that owners understand their pets' needs during each phase. Depending on your kitten's age when the two of you meet, one or two of these stages may have already passed by the time you take your new pet home, but remember that kittenhood lasts for at least a year—or even longer for some breeds.

THE FIRST EIGHT TO TWELVE WEEKS

If you are lucky enough to find your kitten before it has been weaned, be sure to spend some time getting acquainted before your pet's homecoming day. Be respectful of your breeder's schedule. Never show up unannounced, for example, but do ask when you may visit to spend some time with your kitten. A good breeder will be glad to know that you are already making your relationship with your pet a priority.

Although your kitten will be eating solid food and becoming more independent from its mother by eight weeks of age, it will be learning valuable skills over the next month or so that it spends with her and its littermates. Playtime with siblings will include stalking behavior stemming from the kittens' hunting instincts. They may also chase, lick, pounce on, swat at, scratch, and even bite one another. Not to worry—it's all in good fun, as well as part of the kittens' education. Important boundaries are established through this practical play.

Your kitten will also be capable of bonding with you while you visit during these last few weeks. Kittens this age can form surprisingly strong attachments. To help make sure you remain familiar to your new pet, especially if you can't visit as often as you'd like, bring a gift or two that can stay with your kitten even after you leave. No, this isn't about bribing your new pet with toys, but rather keeping something with your scent on it nearby. A small, soft blanket is ideal.

You can use the time that your kitten must remain with its feline family to prepare for your pet's homecoming. After waiting several weeks for your new pet to be old enough to come home with you, the last thing you will want to do is run to the store when you could be home playing with your precious new kitten.

Kitten Fact

Before making a list of possible names for your kitten, consider limiting your options to names with just one or two syllables. Kittens can learn shorter names more quickly than longer, more complicated ones.

FROM THREE TO SIX MONTHS

A kitten between the ages of three and six months of age is a lot like a toddler. A constant toggling between autonomy and dependence is taking place at this time. One minute your kitten will insist on exploring on its own; the next it will want your undivided attention. To make the most of continued bonding with your pet, take advantage of every play opportunity that presents itself. Your kitten may not initiate playtime the same way a puppy might, but by observing your pet carefully, you will see when it is feeling especially frisky or when something catches its eye.

You will also notice during this time period that your pet's energy is on the rise. Your kitten's activity level will continue to increase as it becomes more and more coordinated and confident. Playing with your pet frequently will help to provide it with much-needed exercise—and also help ensure that your kitten sleeps when you do. Kittens that don't get enough exercise during the day can develop a hard-to-break habit of running around the house at night.

The final booster shots for your kitten's vaccinations will also take place during this time period. You and your vet will determine which vaccines are right for your pet. Even the timetable for certain vaccines can vary depending on where you live. A rabies shot, for example, is required by law in most US states, but you may need to revaccinate in one year or three depending on your exact location.

Most female cats come into their first heat by the time they are six months old. For this reason your veterinarian may suggest spaying your kitten before it reaches this age. If you purchased your cat from a breeder, check your sales agreement to see if a certain age was specified for this simple operation. Since male kittens don't reach puberty until about nine months of age, you will need to wait just a little longer to have it neutered. Cats that are spayed or neutered before they reach sexual maturity face a smaller risk of certain cancers, but sterilization is a smart step even if your kitten is past this milestone.

FROM SIX TO NINE MONTHS

If your kitten had remained with its littermates, it would likely be working toward establishing its position in the social hierarchy by now. Without these feline family members to outrank, your kitten may start throwing its weight around with you instead. This stage may be compared to that of the unruly human teenager—instead of talking back, your kitten might scratch the furniture or destroy your belongings to demonstrate its dominance to you. It may even scratch or bite you or other household members to push limits.

Owners must be careful to assert just the right amount of discipline during this phase. Your kitten is testing you to see what it can and cannot get away with. Your job is to let your pet know in a calm yet certain manner that you are the one in charge. Avoid harmful reactions like yelling if your kitten acts out, and never under any circumstance strike your pet. Behaviors like these will destroy all the work you have put into building your relationship with your kitten. You could also hurt your tiny pet.

If your pet is destroying your possessions, remove its access to these items. If your kitten bites or scratches you, quickly say, "No!" and walk away. Although it may be difficult to temper your reaction, try not to cry out even if you feel some pain. Some

animals perceive a reaction like this as engagement—or worse, as a victory. You simply want to say the word no and then ignore your pet.

It is important that everyone in your household has the same rules for the kitten. If even one person plays rough with your pet, it won't understand why aggressive play is allowed with one family member but not another. Make sure all the family is working toward the same goals with your pet.

 Kitten Fact

A kitten's teeth are sharper than an adult cat's.

NINE MONTHS TO ONE YEAR

Although your kitten still has some growing to do, it will be well on its way to becoming an adult cat by the time it reaches nine months of age. Like that human teenager, though, your cat may look all grown up long before it actually is. Some breeds still have a way to go in terms of both physical growth and mental maturity by this age. The Ragdoll and Maine Coon, for instance, won't reach adulthood until they are between three and five years old. No matter what your cat's breed, however, this stage is when those adult characteristics will begin emerging.

Ch-Ch-Ch-Changes

Depending on its breed, your kitten may go through different kinds of physical changes during the first year of its life. The Scottish Fold, a breed known for its folded ears, may or may not have ears that fall forward. All Scottish Fold kittens are born with straight ears. Even breeders can't tell which ones will develop the trademark characteristic until a litter is about three to four weeks old. Identifying show-quality kittens takes much longer—up to 12 weeks. A Selkirk Rex may not fully develop its curly coat until it is about eight months old. And a York Chocolate is born with a light brown coat that becomes considerably richer with age.

As your cat's metabolism begins to slow down, its body will start to fill out. Your pet might enter this phase a lanky-looking kitten, but over the next few months it will grow into a strong, well-muscled animal. It is important to pay attention to your kitten's weight during this phase, as some kittens can gain too much as their energy level stabilizes. If you didn't reduce your pet to twice daily feedings when it reached six months, you definitely need to make this move now—or risk obesity by early adulthood.

You may notice more self-grooming during this period, as well as increased shedding. Minimize the spread of cat hair on furniture and clothing by providing your cat with its own bed, if you haven't already. Even if your cat sleeps with you, it may appreciate having a place of its own for daytime naps.

If it seems like your kitten is losing interest in playing, it may be time to conduct an inventory of its toys. Items that once mesmerized your pet may not hold the same fascination for your older kitten. You may be able to get it to play more by making a quick trip to the pet supply store. You might also consider investing in more challenging items, like treat mazes and other puzzle games. Also, be sure to start rotating your pet's toys to make sure its interest level in play remains high.

BEHIND THE SIGNS

Nipping It in the Bud

Neutering a male kitten can help prevent certain unpleasant behaviors from emerging. Neutered cats are far less aggressive than those that are left intact. Some male kittens also begin marking, or spraying urine, when they enter their adolescence. This behavior is largely performed in an effort to establish territory and dominance over the other members of the household, feline or otherwise. Sterilization often reduces or eliminates this tendency entirely, however. By taking a proactive approach and having your kitten neutered before any problems arise, you may never have to deal with them at all.

No. 12 KITTEN SENSES

THE WORLD ACCORDING TO YOUR FELINE

*Y*ou've probably heard more than one person insist that companion animals like cats and dogs are colorblind, that they only see black and white. Likewise, it is often said that cats can hear—and even smell—another cat or dog from a mile away. And who hasn't seen a story on the news about animals that seem to know ahead of time when their owners are coming home? These seemingly clairvoyant pets make excellent human-interest stories, but is there any truth to them? Are any of the things we've been told about animals' senses actually true?

MORE THAN MEETS THE EYE

Contrary to popular belief, your kitten is not colorblind. Not completely, anyway. Your kitty can indeed see colors, but its range is far more limited than your own. Your pet can distinguish shades of blue and green, but it cannot see red. Wild cats, which hunt at dusk, have no need to see most colors. What at first may seem like a disadvantage is just evolution focusing more on the senses that a cat utilizes for its survival.

One thing your kitten can do far better than you can is seeing at night. Walk into a pitch-dark room with your kitten, and you may become disoriented for a minute or two as your eyes adjust to the lack of light. While you are running your hands along the wall for a light switch, though, your pet is probably moving along without skipping a beat.

In addition to being able to see well in low light, your kitten's eyes can also follow moving objects with impressive accuracy. This too is a useful ability for hunting. Kittens do not see in great detail from far away; cats fall on the nearsighted end of the vision spectrum, but they are excellent at judging distance within a six-foot (1.8 meter) vicinity. As a kitten becomes more coordinated, this skill helps it to know when it's safe to jump from one object to another.

BEHIND THE SIGNS

The Dark Truth

Cats are said to be able to see in total darkness, but this is actually an old wives' tale. No animal can see in the complete absence of light. Your kitten can, however, see in about one-fifth the amount of the light that people or other animals need to perceive visual images. In the wild this ability is useful both for hunting and avoiding predators. Your kitten is more likely to use this skill for less exciting tasks, like finding its litter box in the middle of the night.

Tuning In

Your kitten isn't just good at hearing faraway sounds. It is also skilled at identifying the direction from which the noises are coming. If you watch your kitten when it is listening to something intently, you will notice that your pet turns its ears in reaction to the sound without moving its head. Your pet can actually rotate each ear up to 180 degrees for this purpose, as well as move both ears in different directions simultaneously.

GLAD TO HEAR IT

When kittens are born, they cannot hear at all, but they more than make up for this temporary deficiency by the time they are about a month old. A kitten's sense of hearing typically emerges around two weeks of age and gradually intensifies. By the time they get their ears, so to speak, they can hear 65,000 cycles per second, which is much better than either people or dogs. A greater portion of your kitten's brain is devoted to processing sounds than either yours or your dog's as well.

Kitten Fact

Kittens often equate negative experiences with scents. For example, when you bring your kitten home from a vet visit, another cat in your household may become agitated when it smells where the two of you have been.

Like their night vision, a wild cat's sharp hearing comes in handy for hunting. A pet kitten, though, uses its ears for more domestic purposes. Those cats that are sitting in the front window as their owners pull into the driveway likely heard their favorite humans coming. Believe it or not, your kitten can hear your car engine from an impressive distance, and it may even be able to tell the difference between your vehicle and your neighbor's. Your pet can also identify the sound of your footsteps as you exit the car and head for the door to your home.

Some kittens enjoy listening to music. If you watch your pet when you play certain songs, you may even notice that your kitten has a preference for a certain genre or artist. If your kitten comes running whenever you play Elton John's "Honky Cat," but makes a B-line for the other end of the house when you play Harry Chapin's "Cat's in the Cradle," your pet may prefer British rock to American, although I recommend giving the Stray Cats' "Stray Cat Strut" a try as well.

Because your pet's ears are so sensitive, though, you should use care when it comes to volume, no matter what songs make your pet's playlist.

IS THAT DINNER I SMELL?

Of all your kitten's senses, its ability to smell is the first to emerge. Long before your kitty could see or hear, it was learning about its environment from the various scents he could detect within it. For example, newborn kittens rely on their sense of smell to find their mother's milk.

Smell is intricately linked to eating for cats. A kitten is far less likely to eat food without a strong aroma. Heating wet food, especially if it has been stored in the refrigerator, can help a finicky eater find its meal more appetizing. Owners must be vigilant of respiratory infections, which can block the nasal passages. If a kitten cannot smell, it may not eat at all, which will only complicate any health problem.

A kitten's nose is only slightly less powerful than a dog's. Your kitten has between 50 and 80 million scent receptors in its tiny nose, whereas a dog has between 200 and 300 million. Rest assured that when you open a can of cat food and your precious pet appears out of nowhere, it was just as likely its scent that alerted your cat to mealtime as the sound of the can opener.

HOW SWEET IT ISN'T

When your kitten encounters a strong scent, it tastes the fragrance almost as much as it smells it. Cats have a special scent receptor in their mouths. Situated just behind its front teeth, the vomeronasal organ is more commonly known as the Jacobson's organ. Your kitten may be sitting beside a screened window when its lips start to curl. Other signs that your pet is experiencing the Flehmen reaction, as it's called, include squinting, flattening the ears to the head, and baring teeth. If you notice these behaviors, especially in conjunction with one another, chances are good that a bird, a squirrel, or another cat is nearby.

This turbo-booster for tasting helps wild cats determine whether certain objects are safe to eat. It is also this ability that has earned cats the reputation of being finicky eaters. Most kittens will readily eat cat food that has just been spooned out of the can, but leftovers are often met with much less enthusiasm, usually because they don't tickle the Jacobson's organ like the freshest food does.

One would think that the Jacobson's organ translates to an acute sense of taste, but this is not at all the case. Your own taste buds outnumber your kitten's 20 to one. Your pet has only about 500 of these nerve endings on that rough little tongue. Your pet also cannot taste anything sweet. The only tastes that kittens can detect are salty, bitter, or sour.

KEEPING IN TOUCH

In the wild, cats rely heavily on the four previously mentioned senses, but for a pet kitten, one could argue that the most important sense of all is touch. It is after all human touch that makes all the difference in those first few weeks of life. Touch also becomes a way for kittens to show affection. Two kittens will often cuddle together for companionship as much as warmth. Some kittens have even been known to snuggle with dogs.

Cats also use their sense of touch in more practical ways. The most sensitive parts of your kitten's body are its whiskers, its

paw pads, and its tongue. The nerve endings at the end of each whisker help a kitten detect the slightest motions in its vicinity. In the wild this ability can be helpful for avoiding predators. Your pet's whiskers also come in handy for moving through small spaces. If you've ever wondered how a cat always seems to know which spaces are just large enough for it to fit through, it is the whiskers that make this keen judgment possible.

Kitten Fact

Your kitten doesn't just have whiskers on its face. It also has them on its lower front legs. These whiskers, which perform a similar job as the others, are located just above your kitten's feet.

I'm Hooked!

Some people say that a cat's tongue feels like sandpaper, but it's actually hundreds of small hook-shaped bumps called papillae that give the tongue its rough sensation. The papillae help your kitten groom itself, as they pull loose fur from your pet's coat much like a comb. The hooks can even be a little too effective at removing that hair, however. The papillae can bring so much fur into the mouth that a kitten can develop hairballs as a result.

No. 13 KITTEN COMMUNICATION

TRANSLATE VOCAL AND BODY LANGUAGE

hat does a kitten say? Although this may seem like a simple question for a preschooler, the answer is far more complex than even many grownups realize. Just the word "meow" alone can mean several different things, depending on the way in which your kitten delivers the catchphrase. (Think "Aloha.") Throw in all the other sounds that a kitten can utter—not to mention the plethora of meanings behind the different types of feline body language—and you begin to see what a multifaceted animal your little kitten actually is.

The problem with having so many forms of expression at your pet's disposal, though, is that without a clear understanding of how it communicates, you could be left in the dark as to what your kitten wants. Even worse than not having a clue what your pet is trying to tell you is misunderstanding the message. Obviously, it won't be the end of the world if you feed your kitten an hour early when it just wanted to be petted, but what if your pet is in pain? A basic understanding of kitten communication can make all the difference in situations like this one.

THE CAT'S MEOW

You may notice that your kitten meows whenever it enters a room. Think of this vocalization as the baseline meow. Your pet is saying hello in the most basic way. While it probably doesn't need anything at this moment, don't forget to respond to your kitten. How would you feel if you walked into a room and greeted someone only to be completely ignored? Respond to your pet's social gesture with a friendly welcome. This is also a great opportunity for working on teaching your kitten its name.

Most other meows indicate one of two things. Either your kitten wants attention, or it's feeding time. In both cases the sound will be similar to the meow your pet uses for greetings, albeit a bit louder or more pronounced. Even if you are short on time, always try to offer your pet a minute or two of your time when it requests it. Doing so will make your kitten feel loved and appreciated—and acknowledged. All of these feelings will help your pet grow into a well-adjusted adult. A kitten that is repeatedly ignored, even when it demands attention, may eventually withdraw and become far less social over time.

Of course, not all meows fall into these three simple categories. Meows can also mean that a kitten wants to play, go outside, or simply make noise. To their owners' chagrin, some kittens are prone to meowing for this last reason late at night. Bear in mind that cats are largely nocturnal animals. Rest assured, though, that you can reduce nighttime vocalization—and general activity—by providing your pet with plenty of exercise during the day.

Other sounds that your kitten may make include chirping, growling, hissing, purring, and yowling. Your pet may make any or all of these sounds, so don't worry if you've never heard your kitten growl, for instance. You may even want to be thankful, as this sound is not a sign of pleasure.

Your kitten's first chirp may catch you off guard. Oftentimes this vocalization occurs when a cat notices a bird or other prey animal from a window. At first you may not even be certain exactly *what* it was you heard your kitten do—it can definitely sound a little odd.

Experts offer up several theories on why the feline species responds to prey in this way. A common explanation is that the kitten is frustrated over being indoors while the object of its hunting affection is

outdoors. Another thought is that chirping is an automatic, almost involuntary reaction to seeing prey. And yet a third theory suggests that the kittens give off the sound as a means of controlling their excitement over the prey's presence. There may be some truth to all three theories, or it could vary by the individual kitten.

Growling and hissing indicate displeasure in one form or another. Put simply, these are warnings. Oftentimes a frightened kitten will growl or hiss at another animal in an attempt to scare it off. Be careful handling your pet if it makes either of these unpleasant sounds. Even if you aren't the one causing your kitten's distress, you could get bitten or scratched while it is caught up in the emotion behind the ugly sounds.

Like meowing, purring is commonly misunderstood. This unique sound can mean a variety of very different things. A kitten may purr as a sign of contentment, or it may purr to calm itself when it feels stressed. It may even purr as a means of letting another animal know that it means no harm. The best way to know what your kitten means when it purrs is to observe it carefully over time. Eventually, you will not only notice that different purrs sound remarkably dissimilar, but also learn what each one means to your specific pet. The speed and volume of the purr, as well as the body language that goes along with it, can all play roles in what your kitten is trying to say through purring.

One might call the yowl the trademark sound of the feline temper tantrum. When your kitten makes this unmistakable noise, it wants something—right now. A yowl may sound a little like an intense meow, but much louder, longer, and dramatic. When a kitten yowls, something is wrong. It may be something that is easily fixable, such as feeding your pet the meal that you missed.

But yowling can also mean that your pet is in pain. Begin by taking a mental inventory of your kitten's daily needs: food, water, a clean litter box . . . If your pet has all of these things, perform a cursory examination. Make sure your kitten isn't bleeding, limping, or otherwise injured.

If you have adopted a kitten that was previously allowed to roam freely outside, the problem could be that it wants to go outdoors. If you have decided against allowing your pet to wander on its own, don't give in to the yowling. You can transition your kitten from being an outdoor pet to tolerating its new indoor status. Your pet's young age alone stacks the odds in your favor with this task, but it may take a little time—and a whole lot of patience for the yowling in the meantime.

 ## Kitten Fact

Scientists have discovered that sounds within the range of the average purr (25 to 150 hertz) have healing properties. Just holding a purring kitten as you recuperate from a pulled muscle or broken bone can actually reduce your recovery time. Even cats themselves benefit from this unusual effect—the species is known for healing faster than dogs from broken bones.

SOUND OFF

Your kitten's breed may affect how vocal it is. The following breeds are known for either being especially vocal or particularly quiet:

VOCAL BREEDS

- Burmese
- Japanese Bobtail
- Maine Coon
- Oriental Shorthair
- Siamese
- Siberian
- Singapura
- Tonkinese
- Turkish Angora

QUIET BREEDS

- American Curl
- Birman
- British Shorthair
- Chartreux
- Cornish Rex
- Havana Brown
- Persian
- Ragdoll
- Russian Blue

If only we could ask our pets what they were feeling, and they could answer in our own language. Pet ownership would certainly be a whole lot easier if this were possible. Alas, while our feline friends are capable of some wondrous things, human speech isn't one of them. If it makes you feel any better, consider the fact that people do have this ability, but even they don't always use it. Some people just aren't good at talking about what's on their minds. Some may simply keep their thoughts to themselves, while others might even lie. You see, even the ability to speak offers no guarantee of good communication. A person's body language is often a better indicator of what he or she is truly feeling at a given moment—and the same can be said for your kitten.

A happy kitten stands out right away. It holds its head and tail high with its ears perked up and its whiskers relaxed. Some kittens wave their tails from side to side when they are feeling especially cheerful. In either case, a content kitty looks relaxed. After a long day—or a spirited exercise session—a happy kitten may curl into a ball or stretch out on its back. It may even rub against you, ready for some more personal attention.

If your pet becomes interested in something, it may move its ears forward a bit. This is a sign of a friendly kitten. Moving the ears forward can also simply be a sign that your kitten is alert, taking in all the sounds of its environment. An inquisitive kitten will hold its whiskers straight out from its face. Its hair will remain smooth and flat, with its tail either still and upright or close to the body. It may also be more vocal, meowing or purring, if it encounters another animal or person.

Try not to look your kitten directly in the eye. While this gesture is a way that people offer their full attention to each other, it holds a different meaning for the feline species. To your kitten, staring can be taken as a threat. For this reason, focus on a different body part—or even a nearby object—whenever you interact with your pet. What matters most is that you speak in a soft and gentle tone, for friendly sounds easily cross the species barrier.

It's usually easy to tell when a cat is unhappy, but it can be a bit trickier to know whether the displeasure is stemming from fear, anger, or sadness. An especially frightened kitten will often make itself scarce mighty quickly, making your job easier, but your pet may also remain completely still when it is scared. Other signs of fear include bulging eyes, ears flat against the head and whiskers back against the face, and raised hair on the back or tail. A fearful kitten may crouch, its muscles tense. Listen for growling or hissing, as these are also common behaviors when fear is present.

Just as your kitten can interpret staring as a threat, it may stare at you or others if it means to act in a threatening manner itself. Like a scared animal, an annoyed kitten's hair may be raised, and its muscles are likely to be tense, but it will swing its tail in a low arc. Sudden whipping of the tail is also a sign of anger. Also like a scared kitten, an irritated one may hiss or even screech as a warning. The more of these signs a kitten is displaying, the smarter it is to give it some space.

WHY THEY DO WHAT THEY DO

*N*o matter how much time we spend playing—and working—with our kittens, these young animals need a certain amount of time to learn how to navigate the world around them. Cats are among the most intelligent species in the world today, but like young humans, it takes them a while to adapt to their surroundings, their owners, and even each other. In the process they often do some comical, perplexing, and downright adorable things. You can play a helpful role in your kitten's development by understanding why it does these things— and responding in the best ways possible.

BASIC INSTINCTS

If you are lucky enough to witness your kitten shortly after its birth, you may be surprised by how active it can be even before it can see, hear, or move much at all. The first two weeks of a kitten's life, in fact, are dedicated primarily to understanding that it is now outside its mother's womb. At two weeks the eyes are usually open, and it is beginning to take in sounds. At this point it is still learning to move its ears in the direction of the noise, however. Those talented ears are going to come in mighty handy later on, but first, your pet must figure out how to use them.

Soon the competition begins. While there may be plenty of mother's milk to go around, this fact won't stop your kitten and the rest of its siblings from jockeying for position in the litter. The most comfortable spot at nursing time is the first reward that comes along with top status. In some ways the effort to outrank littermates by literally pushing them aside might be seen as a move toward independence, but at this age your kitten's entire world still revolves around its mother and siblings. Remaining physically close to its biological family is laying the foundation for future learning and bonding with you and everyone else your pet will encounter.

The best breeders know that it is important to strike a happy medium at this point. While most of the kittens' time should be spent with their mother, they also need to be handled by people if they are going to make good pets. The best way to accomplish this goal is handling the kittens without taking them away from their whelping box. Kittens that are handled gently by humans for between 15 and 40 minutes each day during the first seven weeks of their lives actually develop larger brains.

By the time a kitten is three weeks old, it can see well enough to recognize its mother by sight. The animal's sense of smell is also emerging as its hearing also becomes more acute. Kittens stumble frequently during their first few weeks, but by the time they are about a month old, most can stand and walk with relative ease. The real fun comes in another week or two, though, when they start learning how to run, pounce on one another, and even bite with their brand-new teeth. You may think you need to intervene when you see your kitten roughhousing with its littermates this way, but it's best to stay out of the fray if possible. They are unlikely to do any real damage to one another, and these crude games are actually helping them learn how to interact—and how *not* to interact—with others.

PLAY—THE WORK OF KITTENS

From eight to 14 weeks of age, kittens play—a lot. This recreational activity can be a great deal of fun for both you and your pet, but it also has a greater purpose. While your kitten is chasing its tail (a common occurrence) and ambushing your ankles whenever the opportunity arises, it is learning what it means to be a cat. Many play behaviors (such as pouncing) actually stem from a kitten's natural hunting instincts. Other behaviors keep the animal—and often everyone around it—entertained as it becomes better coordinated.

As your kitten moves from living with its littermates to becoming a member of your household, it may continue some of the behaviors that it began in the whelping box. While some actions may be harmless, others are not. Attempting to bite or scratch human family members may be completely normal. After all, play aggression, as it is called, is how your kitten tested limits with its siblings. It is your job, though, to teach your kitten that these kinds of behaviors are not allowed in its new setting.

Your kitten will continue to play with great intensity throughout its first year, but the specific ways that it amuses itself will gradually change over time. Even older kittens will remain curious, even outright mischievous, but as they move from adolescence into young adulthood, most kittens begin to dial back the pace just a bit. You may notice that your pet plays for shorter time periods, or that it isn't quite as rough when roughhousing. Both changes are completely normal.

ALL ABOUT SCENTING

Once your kitten gets used to your household routine, you may notice that your pet likes to flip or roll onto its back in front of you—often at the same times each day. Your kitten may even make a point of rolling into a specific object, such as your shoes while you are getting dressed in the morning. Some owners assume that their pets are asking to be petted when they perform this common ritual. And this may be what your kitten wants, but don't be too quick to assume so.

Many times kittens roll over this way to spread their own scent onto their surroundings or to feel the carpet rubbing against their bodies. Some kittens tend to roll over whenever they spot a ray of warm sunlight. If your kitten welcomes tummy rubs at these times, by all means continue performing them, but since this behavior is frequently accompanied by a devilish mood, be warned that you could

 Kitten Fact

Some behaviors are linked to a kitten's breed. Persians, Ragdolls, and Scottish Folds will happily sit on their owners' laps for extended periods of time. Calm in nature, they relish being still. An Abyssinian, Devon Rex, or Siamese kitten, on the other hand, is much more likely to follow its owner from room to room. These active breeds prefer to be in on as much household action as possible. All of these breeds enjoy human companionship, just in different ways.

receive a playful nip or slap. Since kittens are creatures of habit, expect to meet again—same time, same place— tomorrow.

Scenting is also the driving force behind another common kitten behavior, called bunting. If your kitten appears to be head-butting you, it may simply be trying to rub its scent on you. A kitten who demonstrates this behavior may go about it one of several different ways. Some kittens like to rub their mouths or the sides of their faces on their owners' shins or ankles. Other kittens tend to get right up in their owners' faces and rub forehead to forehead. Mind you, either type of bunting can be performed with

angelic softness or brute force. It truly depends on the individual animal's personality.

Many kittens spend at least some time each day kneading. This behavior, which involves pushing the paws into another animal, person, or inanimate object, typically begins when kittens are nursing. They knead their mother as a way to stimulate the release of milk. Weaned kittens often knead soft objects, such as blankets, pillows, or furniture cushions. While kneading feels good, it too is a way by which kittens spread their scent. All this scenting is a means of claiming territory, particularly in a household with multiple pets.

HOW RUDE!

During the first few weeks after your kitten joins your household, you will witness many endearing behaviors in your new pet. The way your pet nudges you when it's time for dinner or the way it curls up for a nap on your lap while you read a book can instantly bring a smile to your face. But other behaviors are a bit less pleasant. Some can even seem outright repulsive.

If you watch your kitten as it interacts with other pets in your household, you will quickly notice that animals have a completely different way of relating to one another than people do. A fellow cat in your household may introduce itself to your new pet by sniffing its face. As long as both parties are at ease, your kitten will likely respond by sniffing back. Soon, though, one of the felines will inevitably offer its back end for sniffing as well. Whether we accept it or not, the simple

fact is that animals get to know one another by sniffing each other's bottoms—and your adorable kitten will be no exception. For animals, this gesture is a way of expressing good will. *I mean you no harm*, if you will.

Most owners accept this behavior when they observe it in their pets, but find it harder to tolerate when their kittens initiate the same behavior with them. If your kitten climbs up and situates its rear quarters directly in front of your face, it is probably trying to tell you that it considers you a friend. Granted, it may be difficult to take the gesture as a compliment, but try to understand that your pet doesn't realize that it is offending you. Simply move your pet out of your face and go on with whatever you were doing. Try not to react negatively, as it could make your kitten feel like it has done something wrong. Even if the gesture was rude, the sentiment was well intended.

SCRATCH THAT

Scratching is a common—and necessary—kitten behavior. When your pet uses its claws in this way, it is shedding the outer sheaths of its nails to keep them healthy. Scratching also exercises important leg muscles and, like so many other feline behaviors, allows a kitten to claim its territory. In addition to leaving its scent behind, though, scratching also leaves a visual mark. Offering your kitten a scratching post can prevent it from destroying your furniture or other possessions. Dealing with this problem can be frustrating, but remember that the behavior itself is healthy. You simply need to direct your pet to a proper outlet.

$\mathcal{N}$o. 15 KITTEN PLAY OR PROBLEM?

UNDERSTANDING SCRATCHING AND BITING

*D*on't allow your kitten to scratch or bite or you. It sounds simple, doesn't it? But this basic advice prompts instant questions among many new cat owners: How do I stop my kitten from scratching me? Why is my pet biting me? Will this phase pass naturally? Let's start with that last one and work our way backward.

Kitten Fact

You can often predict play-aggression by assessing a kitten's body language. Before the animal makes its first move, it will whip its tail back and forth. With its ears flattened against its head, the kitten will then attack its prey by pouncing on it.

Many new owners are completely unprepared for the reality of normal kitten behavior. While your kitty may indeed be loving most of the time, at certain moments it may act aggressively toward you and other household members. Fortunately, most kittens outgrow the aggressive phase, but in the meantime it's important to know the difference between normal limit pushing and an intensifying behavior problem.

LITTLE ANGEL?

You probably have a heartwarming image in your head of how your new kitten will act once you bring it home. Especially if you have been waiting weeks for your pet's homecoming, it can be easy to idealize those first few weeks with your loving little ball of fur. You may plan to spend plenty of quality time cuddling with your new companion, or watching it play nicely with its toys or your other pets. And at first everything may seem to be going according to your plan, but soon you might start to wonder if you indeed brought home a kitten—or a baby dragon with fur.

Keep Them Busy

One of the most effective ways you can stop destructive chewing is making sure your kitten is getting sufficient play time. Many kittens chew as a way of coping with teething pain, but some chew on virtually anything in sight simply out of boredom. Kittens have loads of energy. They need mental and physical stimulation. They also need interaction. By playing with your kitten twice a day for 10 to 15 minutes each time, you may find that your unruly pet is more interested in napping than in chewing once playtime is over.

SCRATCHING THE SURFACE

Your kitten's tendency to use its claws is as natural to your pet as breathing. From the time it was playing with its littermates, your kitten has used its claws to practice its hunting skills. Although housecats were domesticated long ago, they retain their prey drive—and the skills that go along with it. As with many types of behavior, this one too is more intense with some breeds than others. A Burmese, for instance, may act more aggressively than a Birman during kittenhood—and through adulthood, for that matter.

 Kitten Fact

If your kitten starts scratching or biting every time someone goes near it, it could be suffering from an illness. Schedule an exam with your veterinarian as soon as possible to rule out a physical cause for the behavior.

Aggression is a strong word. Although it's hard to view scratching as a playful gesture when you're the one nursing wounds with antiseptic spray, it is essential to understand that most kittens mean no harm when they use their claws. Perhaps a more fitting phrase for the scratching done by the majority of young cats is aggressive play, or mock-aggression. These little kitties are pretending to hunt, pretending to be wild, and pretending to be fierce. Sometimes they are simply a bit more convincing in this pretense than they realize.

The key word in aggressive play, however, is *play*. All the kitten wants is to have fun. Its siblings responded to being scratched or bitten by retaliating. While this reaction was effective for its feline family members, it is neither a practical nor an acceptable option for human family members. Plus, there's that not-so-tiny matter of needing to teach your kitten instead of merely averting its scratching one attack at a time. Your best defense is to shut down playtime immediately when your pet uses its claws. Harsh reactions like yelling might frighten your kitten, and a loud shriek may give your pet the impression that it's part of the game. But a quick "No!" followed by walking away will let your kitten know that scratching equals the end of the fun.

You can always resume playing with your pet once it has had a chance to absorb this lesson. In fact, returning to play after a short break is the best thing you can do, as it shows your kitten that you are more than willing to participate in recreational activity—as long as it keeps its claws to itself. If your pet scratches you again, again shut things down. This time, though, wait a bit longer before resuming playtime. Eventually, your kitten will get the point.

You may be tempted to allow rough play while your kitten is still young. After all, your tiny pet may not even be coordinated or powerful enough to hurt you yet, but make no mistake, allowing your kitten to play rough now will only instill the very behaviors that you will be desperate to correct in the near future when your tiny pet becomes defter and stronger. And the more deeply ingrained a behavior becomes, the more difficult it will be to reverse.

TAKING A BITE OUT OF AGGRESSIVE PLAY

Kittens bite for the same reasons that they scratch. This behavior begins during the first few weeks of life, and it is even valuable for teaching social cues. When a kitten bites a littermate and the sibling bites back, the kitten learns that biting hurts. In this way, spending time with littermates is a lot like attending preschool and learning how to play nicely with others. But kittens also use their teeth for another reason—they are sore from the teething process.

If your kitten pounces on you and bites you quickly, chances are good that it is practicing predatory behavior. If your pet bites when it *isn't* acting out a make-believe hunting scenario, the reason could be teething. Kittens that play-bite usually do so when they are engaged in play. A kitten that bites in a relaxed position with no other signs of aggression—or even mock-aggression—is likely trying to make its mouth feel better.

The problem with allowing this behavior is that it can become a habit for your kitten. Many kittens that are permitted to bite when they are young continue biting long after their adult teeth have come in. And as with scratching, those bites can hurt a whole lot more as your pet grows bigger and stronger. Even now, your kitten's razor-sharp teeth can inflict some nasty puncture wounds, so if you prefer not to go through that antiseptic spray like water, treat all bites the same—with zero tolerance.

Consistency is essential for teaching your kitten proper behavior. Be it biting, scratching, or any other unpleasant behavior, you mustn't shut it down one day and then tolerate it the next. It may be fun to watch your tiny kitten act like a wild cat, but you will only confuse your pet by indulging bad behavior.

Playing It Safe

If your kitten has begun playing aggressively and you have other pets in the household, it may be wise to keep them separated until this phase passes. Two kittens the same age are unlikely to do each other any harm, but a bigger pet could hurt your kitten by merely defending itself. A large dog or an older cat may have little tolerance for a kitten that scratches or bites. For the safety of all involved, keep a close eye on any interactions you do allow between your pets, and split them up at the first sign of an altercation to prevent it from escalating into a dangerous situation.

If your kitten is play-biting, respond the same way you would to scratching. Offer a firm yet quick "No!" and end the play session at once. If your kitten appears to be biting in reaction to teething pain, however, you will need to find a different strategy. Chew toys can be helpful for diverting your kitten's incisors away from you. Many pet owners are surprised to learn that chew toys are available for cats as well as dogs. They are available in a wide assortment of styles and textures. Some can even be chilled to make your kitten's tender gums and teeth feel better.

Toys should be available to your kitten at all times. You can help maintain your pet's interest in these items by rotating them from time to time. Even a bright and colorful teething ring bent into the shape of a bird can seem old and boring to your kitten if it's the only plaything you ever offer your pet. By keeping a fresh assortment of toys at your kitten's disposal, you may also prevent your pet from sinking its teeth into your furniture and other possessions. A kitten that has no toys is much more likely to gnaw on these items instead.

FIRST, ALLOW NO HARM

In some cases, neither mock-aggression nor teething is behind a kitten's tendency to lash out through scratching or biting. If your cat has a reason to fear a person or fellow pet in your household, it could react by using its teeth or claws to defend itself. Young kids or poorly behaved dogs can pose a particular threat to a small kitten. Maybe your child doesn't even realize that he or she is hurting the kitten when picking it up, or perhaps the dog has been stepping on the kitten's tail accidentally. Whatever the cause, if your cat is acting out in self-defense, it is your duty as its owner to address the problem and protect your kitten from further harm. If you don't, your pet's long-term temperament could be negatively affected by the experience.

LET'S TALK LITTER BOX AND SCRATCHING POST

*B*efore they become cat owners, many people mistakenly assume that kittens train themselves. Whether the task at hand is housetraining or using a scratching post, they assume that all felines simply know what to do from the start. For this reason, many also think that buying or adopting a kitten is easier than getting a new puppy. While it's true that young cats model many of their behaviors after their mothers or littermates, and later after older cats within their households, kittens do not intrinsically know which actions they should—or shouldn't—imitate. The fact is that all kittens need a certain amount of training.

The good news is that most kittens are quick studies. And the even better news is that these intelligent little felines are steadfast creatures of habit. Once your kitten starts performing a particular desired behavior regularly, all you have to do is make it easy for your pet to keep performing it. In the meantime, though, some guidance will be necessary.

LITTER BOX TRAINING

One of the reasons it is so easy to teach a kitten to use a litter box is that cats are instinctually clean animals. By the time you bring your kitten home, it is already washing itself from head to tail several times a day. By the time it reaches adulthood, your pet will dedicate more than half its waking hours to this self-grooming task. Your kitten doesn't just want its own body to be clean; it also wants its environment to be unsoiled. This doesn't mean, however, that an untrained kitty won't use a corner of your home—or even a houseplant—as a makeshift bathroom.

Your kitten probably began the housetraining process by observing its mother's behavior. If you have other cats in your home, your pet may even continue learning by watching and imitating them, but you must set your kitten up for success. Provide your new pet with a litter box, even if the resident cat's box is a large one. Every cat within the household should have a litter box. You may find that your cats use whichever box is convenient at a given moment, or each animal might claim a certain box for itself. But they must have at least one box per cat.

Ideally, you should also provide your pets with one extra litter box. Doing so helps ensure that they always have a clean place to relieve themselves—therefore lessening the chance of house soiling. Do not place all the boxes in the same room, though. Even if one box is perfectly clean, a kitten may turn around as soon as its nose detects that a nearby box has been used recently.

Show your kitten the location of the litter box as soon as you bring it home. To encourage your kitten to use the box, place your pet inside it shortly after meals, naps, and play sessions. You can also transport your pet to the box promptly if you notice pre-elimination behaviors such as sniffing around or crouching. As soon as you catch your kitten using the box, praise it effusively. You may even offer your pet an edible treat if you like. Positive rewards, whether verbal or more tangible, reinforce behavior.

If your kitten has an accident—and it may—don't scold your pet. By the time you discover the misdeed, your kitten will have moved on to something else. And it won't understand why you are upset. Instead, move solid waste to the litter box to help your pet understand that this is the proper spot for eliminating. Don't leave the waste in the box too long, though, as you must remember that kittens prefer a clean environment for doing their business. For this reason it is also essential that all boxes within the home be cleaned regularly, ideally every day.

Some kittens are especially picky when it comes to their litter boxes. Perhaps your pet prefers a box with a cover, or maybe it will only use a box without a lid. It could be that your kitten became accustomed to the type of litter box its mother used,

 Kitten Fact

An especially young kitten may need a litter box with shorter sides. If your pet has trouble climbing into or out of its box, it may begin eliminating elsewhere.

Kitten Fact

It is a myth that mother cats train their kittens to use a litter box. Although kittens may certainly learn about housetraining by watching their mother's example, the mother cat makes no effort at encouraging her young to imitate her behavior.

Stop and Smell the Litter

When selecting a brand of litter for your kitten, you may be tempted to choose one with a pleasant scent. What smells good to you, though, may be an aversion to your tiny pet. Remember that your kitten's nose is much more sensitive than your own. And enclosed litter boxes will only intensify the odor. If your kitten finds the scent too strong, it may even refuse to use the litter box as a way of avoiding the smell. This situation could cause your pet to soil your home or hold its bladder. In time the latter case could lead to serious health problems for your kitten.

or your pet may simply like its privacy. Other kittens are more easygoing about the box specifics, but will only eliminate in a particular type of litter. Finding out what your pet is partial to can take some time, as well as a bit of trial and error. Once you identify your kitten's preferences, though, it is smart to do your best to accommodate them.

While accidents can definitely happen, housetraining regression in kittens is relatively rare. If your pet begins urinating outside of its litter box after being reliably trained for weeks or months, a physical problem could be the cause. Oftentimes a urinary tract infection is to blame in this situation.

THE SCRATCHING POST

Teaching a kitten to use a litter box may strike you as a higher priority than training it to use a scratching post, but in many ways these are equally important tasks. If your kitten is prone to clawing, your furniture, window treatments, and even bedding could be at risk unless you quickly show your pet the correct spot for scratching.

Buying your kitten a scratching post is just as important as making sure your pet has food bowls and a litter box. You cannot teach a kitten not to scratch. And truly, there is no need to try to stop this healthy activity. You can, however, provide your pet with an outlet for this normal feline behavior.

While your kitty is in scratching-post training, you must protect your belongings from its claws. Once your pet has clawed—or even tried to claw—a piece of furniture, treat it with an antiscratch spray. This product will remove the odor left behind by your pet, which can encourage it to revisit the scene of the crime. Cover furniture and other high-risk items with old blankets or towels before your kitten can get to them. Then, begin watching your pet.

When you can't supervise your kitten, contain it in one room. Place the scratching post in that room as well, and reward your pet each time you notice it using the item. As with litter-box training, you may choose to use verbal praise, edible treats, or both. What matters is that you reinforce the behavior. Remember, your little creature of habit will soon find comfort in revisiting the scratching post, even without a reward.

 Kitten Fact

Some cat owners recommend rubbing catnip on a scratching post to encourage a kitty to use it, but this move could prove to be futile with a young kitten. Cats younger than six months of age often have no reaction to catnip whatsoever.

DECLAWING: WHAT THE EXPERTS SAY

A 2013 documentary titled *The Paw Project* addresses the issue of declawing cats and kittens as a means of preventing them from scratching. While the practice is illegal in the United Kingdom and many other nations, this surgical procedure, which amputates both claw and bone, is still allowed in the United States—although both the Humane Society of the United States and the American Society for Prevention of Cruelty to Animals strongly discourage it. Many veterinarians compare the procedure to cutting a person's fingers off down to the first joint. Proponents maintain that recovery from the operation is usually quick, but studies indicate that complications occur in between 25 to 50 percent of animals. These problems include pain, hemorrhaging, and regrowth. Still, some vets insist that declawing is preferable to homelessness when a cat's scratching becomes a deal breaker for continued ownership. Dr. Jennifer Conrad, who directed the documentary, thinks that many vets are willing to keep performing the surgery to add to their bottom line. "We have let it happen here and no one has challenged it," she tells *National Geographic Daily News*, but she is optimistic that *The Paw Project* will make owners reconsider the practice. "I hope that Americans and Canadians begin to really question what's right for their animals."

MISBEHAVING KITTY

HOW YOUR ACTIONS ENCOURAGE OR DETER BEHAVIOR

hen kittens misbehave, owners become frustrated. It's certainly not easy to come home to find that your pet has turned the fabric on your newly reupholstered sofa into fringe—or that it has mistaken the corner of your dining room carpet for its litter box. Even smaller transgressions, such as walking on kitchen counters or chewing on plastic bags, can be more than a little annoying. It's easy to blame your pet for these and other unpleasant behaviors, but a more effective strategy is to consider whether another party might share in the responsibility. No, I'm not suggesting that your kitty has an accomplice. It's important to look at the role you may be playing in your pet's poor choices.

HELPING KITTY LEARN

Certainly, you would never intentionally encourage your kitten to behave badly. You also wouldn't purposely set your pet up for failure. Without realizing it, however, many owners react to unpleasant discoveries like clawed furniture in ways that make it almost impossible for their pets to avoid making the same mistakes over and over again. The key to turning the tide is stopping this vicious cycle and making a conscious effort to deter poor behaviors instead of subconsciously reinforcing them.

The smartest owners quickly realize that the first step to correcting unwanted behaviors is looking at each situation from their pet's perspective. It is also vital to be able to tell the difference between the things that you can and

cannot change about your kitten. You cannot stop your kitty from using its claws, for instance. The behavior is as natural to your pet as scratching an itch is to you. But you *can* offer your kitten appropriate alternatives to destroying your possessions. Owners who respond to bad behavior by scolding and offering no such alternatives are just as responsible for their pets' misbehavior as their animals themselves—perhaps even more so.

 Kitten Fact

In general the feline species is not a fan of change. For this important reason, try not to move your kitten's litter box unless it is absolutely necessary. Doing so could lead to housetraining regression.

START WITH THE POSITIVE

The best way to ensure that your kitten behaves well is by rewarding good behavior whenever it happens naturally. You will be amazed by how much your pet is already doing right if you just take the time to look for these successes. Chances are good that your kitten will encounter some stumbling blocks on the way to becoming a well-mannered adult, but your pet will also master many things easily. Reward your kitten for as many of them as possible!

Using a proactive approach to training is much easier than correcting undesirable behaviors later. Buy your kitten a scratching post *before* it can seek out another object in your home for this purpose. You will feel much less pressure when teaching your pet how to use this item if your kitten hasn't already begun destroying your belongings. That lower stress level will come through to your pet. Animals are astonishingly skilled at picking up on our moods. Likewise,

before your kitten can develop a habit of jumping onto your counters, offer your pet a treat when it stays on the floor while you're making dinner or cleaning up afterward.

Sometimes setting your pet up for success will require adjusting your own habits. If you don't leave plastic bags lying around, your kitten cannot chew them. Moreover, your pet also can't be hurt by them. Living with another species isn't always easy, but some compromises are well worth making. If you haven't been able to find a solution to another behavior problem, ask yourself if there is any middle ground where you can meet your pet. Not wanting your pet on your kitchen counters is completely reasonable, but is it truly necessary for your kitten to stay off all your furniture? Only you know which concessions you are willing to make, but sometimes a small effort can turn a bigger problem into a more manageable one.

LIMIT BAD CHOICES

Offering constructive outlets for your kitten's natural instincts, like scratching, is a great start to getting your pet to behave the way you want. In many cases, though, this single step won't be enough. Leaving your furniture unprotected after your pet has already shown an interest in sinking its claws into it is akin to leaving the keys in your car after a thief has already tried to steal it. Protecting a larger item like a sofa will require antiscratch spray and covers, but for a smaller piece of furniture, like a chair or ottoman, consider moving the item until your kitten's training is complete. You needn't move the piece to the attic—just out of the rooms your kitten frequents.

You won't just be protecting your own assets with these steps. You will also be protecting your pet from performing the undesired behavior a second time. For this reason it is especially important that you don't make a rookie mistake. Once a kitten has clawed an object, some owners see no reason to block the animal's access to it. After all, the chair is already ruined, right? While it may be true that another scratch won't hurt your chair's appearance any further, allowing your kitten to keep scratching it will only confuse your pet. How can you expect your kitten to understand that it's OK to scratch the club chair but not the chaise? The simple answer is that you can't.

This is an example of expecting your feline companion to think like a human. Cats are intelligent creatures, but they cannot tell the difference between two pieces of furniture. Many of them do, however, have an uncanny knack for attacking the most expensive piece in the room—so it might be wise to move any extremely valuable or sentimental items, whether your kitty has made its way to them or not, until you have corrected a scratching problem.

The "Terrible Twos"

A human toddler enters the so-called "terrible twos" right around 24 months of age, just like the name of this ornery phase implies. Your kitten will also go through a terrible stage, but for your pet this period will be around four to five months of age. For a little while, it may seem like that sweet little kitten you brought home has been possessed by a Tasmanian devil. Your pet may get into everything and act out for seemingly no reason at all. Luckily, this stage lasts for far less time for kittens than it does for children. Your kitten is bound to act up a bit from time to time until it reaches adulthood, but the worst should be over in about a month.

DON'T BE AN ENABLER

Sometimes a small problem can grow into a bigger one surprisingly quickly. Perhaps your kitten is a bit of a picky eater. If this is the case, you may see it as your job to find a food that your pet will eat. While technically this may be true, it is very important that you don't let your kitten in on the secret, for your pet might just take advantage of your good intentions—if you let it. If your pet refuses to eat, how you respond can make all the difference in whether your kitten becomes a truly finicky eater or just a master manipulator.

Some owners worry that if a kitten misses a meal, it won't get enough nourishment. They may even fear that the kitten will get sick. Rest assured that one missed meal will not harm your pet, especially considering that kittens eat more often than adult cats. Missing breakfast isn't a cause for concern. If this happens, your pet will almost certainly eat its lunchtime serving. The bigger problem comes when owners start jumping through hoops the moment a kitten refuses a particular food. If the first thing you do when your pet won't eat is open a new can, the only thing you will accomplish is teaching your kitten not to take your first offer.

Some kittens are indeed finicky, but being too quick to give in to them can exacerbate the problem. If your kitten is used to being served several choices at every meal, it probably won't react well to a sudden take-it-or-leave-it approach. If you want to nip the problem in the bud before it gets even worse, however, you must be strong. No matter how much your kitten meows or yowls, wait until your pet's next scheduled feeding to offer food again.

Sometimes a kitten genuinely doesn't like a certain food. If your kitten detests salmon, for instance, but likes every other food you have offered it, there is no harm in eliminating salmon from the shopping list. Reasonable likes and dislikes are not problems, but if your kitten's preferences appear to change with its mood and you accommodate this behavior, get ready to figure a whole lot of wasted food into your budget. Do make sure that your pet doesn't go too long without eating, though. A true loss of appetite can be a sign of a health problem.

Also bear in mind that enabling isn't limited to feeding. You can make any behavior problem that your kitten develops worse by making it easy for your pet to continue it. Remember, one of the most basic principles of training is that if you make it easy for your kitten to repeat a behavior, your pet is sure to do just that. Unfortunately, this is a rule that works both ways. You can instill bad behavior just as easily as good if you aren't careful.

BEHIND THE SIGNS

Crash! Bam! Boom!

As your kitten works on practicing its predatory skills, one of the casualties could be your breakables. And if you come running each time you hear a crash, your pet may even decide that this new pastime makes a fun game. Your best defense is either placing your delicate objects in a display case where your pet can't reach them or packing them away for a while. Until you can make space in your attic, keep your kitten busy whenever possible, and make sure you stick to your pet's feeding schedule. A bored or hungry kitten is more likely to act out, but a tired one will sleep when you can't keep an eye on it.

THE UP-ALL-NIGHT KITTY

SAY "GOODNIGHT" (AND MEAN IT)

The youngest members of nearly every species always seem to have a love-hate relationship with bedtime. And kittens are no exception to this unwritten rule. For a species that is known for sleeping more than twice as much as people, cats have an incredible amount of energy in their waking hours. So it can be especially frustrating when your kitten wants to make its most active period the hours that you need to spend sleeping. You may worry that your tiny housemate will get itself into trouble while roaming around your home solo, or your pet might be making a whole lot of noise at this time, making it impossible for you to get your own rest.

Whatever habits your kitten has developed during the overnight hours, like other bad habits, it is best that you correct them now before they become too deeply ingrained. You may wonder if you can actually stop a kitten from staying up all night. Short of crating your kitten or closing it into your bedroom when you go to sleep, how do you even go about it? And won't your pet protest—perhaps loudly—if you try either of those approaches? It can be pretty tough to make a kitten lie down and go to sleep on command, but you can train your new pet to rest—quietly—when you do.

LONG NIGHTS, SHORT MORNINGS

Whether your kitten is a night owl or not, you will quickly notice that your pet has a definite plan for the start of its day. It may enjoy getting up with you and eating its breakfast while you sip your coffee, but before you make it into the shower, your sleepy kitty probably begins its morning nap. If you need to be at work or another place each morning, preventing your kitten from sleeping all day may be virtually impossible. If you don't have to leave right away, however, you might want to consider some limit setting. Don't prevent your pet from napping altogether. Your kitten needs its rest to be healthy, but it also needs to eat, drink, play, exercise, use the litter box, and get to know its new home.

The less your pet sleeps during the day and the more active it is during daylight hours, the less interested it will be in roaming your digs or putting on a show at night. If you work outside the home,

perhaps another household member can make sure that your kitten is getting plenty of play and exercise when the sun is up. Maybe your kids get home from school in the afternoon several hours before you make it home yourself. If so, make a point of asking them to play with your pet for a little while before starting their homework or doing their chores. (It probably won't take much convincing.)

If no one will be home until the early evening, utilize some of the time before or after dinner (or both) for kitten playtime. You may have the best luck at catching your kitten's interest at this time, as kittens naturally become more active at twilight. Especially if your pet has been sleeping most of the day, it will have plenty of energy for chasing its toys by now. Your job is to expend that energy so your kitten wants to sleep at bedtime.

 Kitten Fact

Kittens sleep a bit more than adult cats. They need between 15 and 18 hours each day. Fully grown cats, on the other hand, usually get just 13 to 16 hours of shuteye.

YOU'RE FEELING VERRRY SLEEPY . . .

The amount of time you spend playing with your kitten during the day and evening can have a significant effect on your pet's energy level at night. You can also use a few tricks to help ensure that your kitten will feel sleepy when you do. Kittens are known for sleeping after they eat. And the bigger the meal, the longer the resting period usually is. For this reason consider offering your pet its main meal just before bedtime.

If your kitten is waking in the night because it is hungry, having a late dinner will also help keep its belly full during the overnight hours. Some owners employ timed feeders to help with this problem. These dispensers release a predetermined amount of food at various intervals throughout the day. If what your kitten wants at two in the morning is a snack to hold it over until breakfast, this item could offer a practical solution. It won't take long for your kitten to learn how the system works. Even if you primarily feed your kitten wet food, you could supplement with dry food at night. Just be sure to adjust your pet's portions to account for this addition.

By far one of the best ways to tire a kitten out during the day is providing it with a friend. Especially if you need to spend most days away from home, having another kitten to play with helps ensure that your pet is getting the companionship, exercise, and stimulation that it needs when you can't provide these important things. Having a contemporary around also means that when your kitten wakes in the night, the two will play together rather than demanding your attention. Of course, they could also wake you if they become too boisterous in their activities.

If your kitten's nighttime activity is making it impossible for you to get the rest you need, you may need to move your pet's sleeping quarters to another part of your home. You needn't place your kitten in its crate for sleeping. Doing so will prevent your pet from using its litter box when nature calls. But making the crate available—with the door open—is a smart idea, especially if your pet likes the enclosure. You might also want to place a pet bed or a comfortable blanket in the room to encourage your kitten to lie down and relax.

The "Midnight Kitten Crazies"

If your kitten starts darting around your home around midnight, don't jump to the conclusion that it will be up all night. Some otherwise perfectly behaved kittens transform into wild banshees as the clock strikes 12. Some kittens happily entertain themselves as they utilize this sudden burst of energy, but others are a little more demanding—waking up their owners by pouncing on their beds. This relatively common occurrence can be frustrating for owners, but luckily, in many cases the animal tires itself out relatively quickly. Most kittens also outgrow the intense "midnight crazies," even if they continue to roam around a bit at night.

IT TAKES TIME

Kittens in particular are prone to nighttime activity, but fortunately for most owners, this species is also capable of adapting to their owners' routines. Breeds that are known for being especially affectionate may value your companionship so much that they adjust to the household routine sooner than others, but all kittens will eventually figure out that they get more of your time and attention if they sleep when you do.

The adjustment won't happen overnight, though. Until your kitten fully settles into your home, it may even feel a little uneasy during those nighttime hours, temporarily increasing the amount of time it spends wandering around in the dark. Try to be patient. You may even expect to lose a little sleep, but do not indulge your kitten by getting up or playing with it. Although

tiring your pet out might seem like a practical strategy to get it back to sleep, save this approach for your preemptive strikes against nighttime activity. You don't want to reward your kitten for being rambunctious when it should be sleeping. You may even find that a middle-of-the-night play session energizes your pet even more, which will just intensify the problem.

As long as you don't reinforce nighttime activity, your kitten will also outgrow this habit to some degree. As your tiny pet needs fewer naps during the day, it will expend more of its energy—instead of recharging its battery—at this time. Some kittens never fully outgrow the tendency to roam around the house at night, but you can lessen the amount of time your pet spends this way. As your pet matures, it may also be much less likely to wake you or get into trouble when it does rise before you do.

 ## Kitten Fact

A kitten's sleeping habits are hardwired into its predatory instincts. Wild cats expend a great deal of energy when they stalk, run down, and pounce on their prey. To prepare, they sleep—a lot. Domestic kitties sleep for long periods of time for similar reasons. They may only be leaping for make-believe mice, but when the opportunity arises, they are ready for the challenge.

 ## Kitten Fact

While many scientists classify cats as nocturnal, most species are actually crepuscular. This means that your kitten will be most active at dawn and dusk.

Risk of Rain

Have you ever noticed that you feel a bit sleepier on cold, rainy days? If so, you're definitely not alone. Many people find that the weather affects their moods and energy levels. Your kitten's vivaciousness may also be noticeably lower when the weather is unpleasant. Even if your kitty doesn't set a single paw outside on these days, it is still more likely to want to sleep for longer periods when the weather is dismal. For this reason it is wise to make a habit of a little extra playtime when you are stuck indoors together. Doing so might help you avoid an extra-early wakeup call in the morning.

A ROOM WITH A VIEW

If you can't seem to get your kitten to sleep through the night, consider giving it something constructive to do with its time that won't keep you awake as well. Window perches are available in a wide assortment of materials and styles. Some even come in the form of clear boxes that sit securely within the window frame to make the kitty feel like it's actually outside. (Don't worry, your pet cannot escape as long as this item is properly installed.) During the day a window perch allows your pet to bask comfortably in the sunshine, but at night it offers a whole new set of sights and sounds for your kitten's sensory pleasure. Remember, your kitty sees best at night, so it will notice things you do not. If your pet makes a raucous noise, many of the nocturnal creatures it spies will fly or scamper off immediately—a fact that won't be lost on your shrewd kitten. Instead of acting out, your pet may soon start crouching down to enjoy the show. It may even fall asleep—eventually.

KITTEN INSTINCTS

INSIGHT INTO YOUR FELINE FRIEND

*B*eing a kitten owner would be so much easier if our feline friends could speak English. Although we may never be able to teach our cats our words, we can understand our pets better by learning as much as we can about their unique combination of instincts. When we know what drives our kittens, their behavior becomes easier to manage, and often our bonds with them strengthen. They might not be able to use our language, but to some degree we can learn theirs.

Feline instincts aren't so much a secret as they are simply misunderstood sometimes. As human beings, we tend to think in terms of our own emotions, motivations, and strategies. When a kitten does something, we automatically go to the first reason we might have for doing something similar. While this thought process may work logically for understanding another person's motives, it often falls short of explaining an animal's behavior.

THANKS, BUT NO THANKS

When you initiate a play session with your kitten and it responds by walking away, your feelings may be a little hurt. You may worry that you have done something to annoy your pet or that you are failing horribly at the bonding process. You might even wonder if the kitten you so painstakingly picked out dislikes like you. First, take a deep breath. Now repeat after me: "Kittens are not puppies." While young dogs—or even many older ones, for that matter—will readily join in nearly any game you suggest, kittens are more independent. They enjoy playing immensely, but they do not depend on people for their entertainment as much as other pets do.

According to British anthrozoologist John Bradshaw, the feline species simply isn't as in tune with human behavior as canines are. He points out that while cats are technically domesticated, they aren't as domesticated as dogs, which have lived with people since the late Stone Age. Bradshaw also points out that while spaying and neutering are responsible steps for pet owners to take, this practice has severely limited domestic animals' evolution. The overwhelming majority of kittens—about 80 percent—are born to stray or feral cats. This statistic means that most cats are inheriting the wilder instincts of these parents.

A kitten's self-sufficient attitude comes at least partially from the species' hunting habits. Early cats hunted alone, and through this process, they became accustomed to being alone much of the time. In general kittens are far more gregarious than their older counterparts, but they too retain some of the aloofness for which the species is known.

If you want to tip the odds in your favor for getting your kitten to play with you, you may need to rely on a few props. Interactive toys like balls that light up can be ideal choices. But don't expect your kitten to play like a puppy. You may be able to teach certain breeds—like the Abyssinian—to play fetch, but most kittens are much more interested in chasing those bright lights than returning the ball to you. If you find that playing ball with your kitten turns into a game of chase, as your pet chases the ball and you chase your pet, consider buying a laser pointer. Your kitten will have just as much fun chasing this little light, but you will always be part of the fun.

Let Your Kitten Come to You

Kittens may not want our attention when we want to give it, but when they want it, they want it immediately. If your pet rejects your offer to play, wait a little while and try again. Try not to take it personally when your pet passes you by. To some degree aloofness is just the nature of the beast. And remember that play isn't the only way that you can bond with your new pet. When your kitten comes to you, gently rub its neck, talk or sing to your pet, or just sit quietly together. Just because your kitten doesn't want to play, that doesn't mean that the two of you can't share a quality moment.

 ## Kitten Fact

Pheromones are as unique as human fingerprints. Just by sniffing, any cat that comes across your kitten's unique scent will be able to tell a lot about your pet—from its gender to how long ago it deposited the scent.

WHAT'S MINE IS MINE, AND WHAT'S YOURS IS MINE

Kittens love rubbing against things. From your furniture to your clothing (usually while you're wearing it), it may seem like your pet is determined to get cat hair all over everything you own. At times your kitten may even dart between your legs, unable to choose which one to rub against first. You might even think that your pet is trying to trip you. Your pet's actual motive for this behavior isn't evil at all, however. Your kitten is claiming you as its own.

The instinct to rub against things or people—and deposit pheromones in the process—is deeply ingrained in the feline species. Although you won't be able to smell the scent left behind by your kitten, other cats will, which is exactly what your kitten is intending. It's your pet's way of saying, "This person is taken. Get your own." Instead of becoming irritated by the fur that goes along with those pheromones, try to take this gesture as a compliment. Picking up a lint brush may be helpful, too.

Like many other feline habits, you can keep this one from leaving excessive pet hair behind by brushing your kitten regularly. If there is something you don't want your kitten touching, put it behind closed doors, as any possessions you leave out will be fair game. Kittens can be ridiculously nosy—I mean, inquisitive. If an item catches your pet's eye, it is probably worthy of a rub.

It's important to understand that sometimes kittens use rubbing as a form of communication. If your pet has been rubbing against you for an extended period of time, it could be that you have forgotten to feed it or fill its water dish. Or your kitten may just be seeking attention. If you have two kittens, you might even witness one of them rubbing against the other. If so, the instigator could be asking to be groomed. This is a common activity among bonded felines.

 Kitten Fact

Kittens that were orphaned or taken away from their mothers too soon may have a strong need to be with their favorite people as much as possible. In fact, these kittens may even suffer from separation anxiety if they are left alone too much.

POCKET PETS AND BIRDS—AKA PREY

Just because your kitten is tiny, do not underestimate the depth of your new pet's hunting instinct. While some cats can live peacefully in households with smaller pets such as gerbils or parakeets, it is impossible to know whether your kitten will be one of them until you get your new pet home. If you already own smaller animals, be sure to keep their cages securely closed unless your kitten is securely contained in its crate. No matter how well behaved your kitten is, it should never be trusted with a rodent, reptile, or bird. In the wild these animals serve as a cat's prey. Even if your kitten has never hunted before, it will recognize these animals as quarry immediately.

If you do observe predatory behavior in your kitten, don't think poorly of your feline companion. More importantly, don't show your displeasure. If a field mouse finds its way into your home, your kitten will likely be the first one to know, informing you of its presence after your pet has caught it. Your beloved kitty isn't a coldblooded killer, but your pet is a naturally skilled hunter. Being admonished for responding to this instinct will probably crush—and at the very least confuse—your pet. Bringing you the carcass is your kitten's way of showing off its accomplishment. It is expecting praise, not punishment.

TOYING WITH PREY

If you come across your kitten just after it has caught a mouse, you may witness a rather tortuous practice. Many kittens will play a cunning game of catch and release with their prey before delivering the final, deadly blow. Experts disagree as to why some felines do this. Some think it comes from a lack of both confidence and experience in hunting. To a young kitten that has never hunted before, prey that can bite back can indeed be intimidating. Others theorize that since domestic cats aren't able to hunt very often, they like to stretch out their fun when an opportunity presents itself.

YOU AND YOUR KITTEN

ESTABLISHING AND STRENGTHENING YOUR BOND

The relationship you develop with your kitten now will be the foundation for the lifelong bond the two of you share. That's quite an intimidating sentence, isn't it? Truly, it can be easy to feel overwhelmed with worry at first. What if you don't get off to the best start with your new pet? What if your kitten has a hard time adjusting to its new home? How will that affect your long-term connection? What if you make mistakes?

Let's get the bad news out of the way—you will make mistakes. And that's OK. Nearly every pet owner does at some point. The good news, though, is that if you make your kitten's well-being your top priority, the inevitable bumps in the road you encounter should be small ones. Bonding with a kitten doesn't happen in a day, or even a week or a month. It is a process that takes time, dedication, and patience. Ideally, you should never stop bonding with your pet.

BOND EARLY AND OFTEN

You don't need to wait until your bring your new kitten home to start the bonding process. You might notice an instant rapport between the two of you the moment you meet. An immediate connection certainly isn't a prerequisite for forging a positive relationship with a pet, but if the two of you hit it off especially well from the very beginning, make the most of the situation. If your kitten isn't ready to come home right away, don't stay away while you wait. Be considerate of the breeder's schedule, but definitely ask when you can return for a visit.

In the digital age we live in, it is easy to rely on methods like email and texting for communicating with one another. Don't get me wrong; there is certainly something to be said for the convenience and fun that all our technology offers.

Your breeder may be kind enough to send you photos or videos of your kitten while you wait for your pet's homecoming day, for example. But when it comes to animals, there is no such thing as an online relationship. As cute and entertaining as these images may be, they are a virtual one-way street. The only way to bond with your kitten is to spend time with it in person.

When you visit the cattery, take along a toy that may interest your kitten and its littermates. Be sure to clear the item with the breeder first, just to make sure that this is OK. You may even choose to leave the toy with your pet and its siblings, again if this is alright with the breeder. The more the kitten sees you, the more familiar it will become with your face, your voice, and your scent—and the more comfortable it will be when the time comes to go home with you.

TIME IT RIGHT

Before you decide to add a kitten to your life, you must be sure that you have enough time for a new pet. If you work long hours or your job is heading into its busiest season, it may be wise to postpone becoming a new pet owner until your schedule calms down. This isn't to say that you can't work and have a kitten at the same time—surely you can, but the timing is paramount. If at all possible, arrange to pick your kitten up when you will be able to spend at least the first few days at home with your pet. Synchronizing the homecoming day with the beginning of a staycation is ideal, but even taking a long weekend is better than trying to introduce your kitten to its new home when you can't be there.

Your kitten's homecoming marks the beginning of an adjustment period for both you and your new pet. Certainly, your kitten's life will be changing the most, but yours too will be different now that you have this special little someone sharing your home. By and large most of the changes for both of you will be positive ones, but you will still need to allow yourselves some time to find your "new normal" nonetheless.

It is important for you to be available to your kitten when it needs you, but it is also vital to give your new pet adequate space. Expecting your kitten to spend every moment with you isn't just unreasonable; it could cause your pet to feel smothered and hide from you. Remember, bonding can't be rushed—or forced. You will do the most good by simply allowing your kitten to get to know you on its own timetable. Respecting your pet's boundaries will help build trust. And this, perhaps more so than anything else, is the best foundation for any relationship.

SLOW DOWN AND TAKE NOTICE

Some owners do everything right in the first few days after a kitten's homecoming, but then they go back to their frantic routines and wonder why they haven't bonded with their new pets the way they had imagined. A big part of getting to know your kitten involves simply slowing down and noticing when your pet is making an effort to get to know you better. Forehead bumping, kneading, and licking are all ways that this species shows affection, but if you are always focused on the next thing on your to-do list, you could easily miss your kitten's attempts at these displays.

If you have to be at work first thing in the morning, consider getting up a little earlier each day. You will hardly notice if the alarm clock rings just 15 minutes sooner, but this small amount of time could mean the difference between having to rush and being able to slow down long enough to enjoy a moment or two with your new pet. You needn't do anything special at this time. Some kittens enjoy just being with their human companions, but a regular play session could certainly set the pace for a fun ritual for the two of you.

You may find that you connect with your new feline friend the most in unexpected moments, but in order for these moments to take place, you must make time for them. Another great way for working owners to make their kittens feel special is to say hello as soon as you get home and to say goodbye whenever you leave. Whether your kitten understands what you are saying or not is irrelevant. Heck, even *what* you say is pretty inconsequential. What matters is that you show your pet that it is important to you.

 ### Kitten Fact

When your kitten licks you, it isn't giving you kisses, at least not the way you might think. Your pet is likely showing you love in another important way, however. Cats often groom one another to show affection, so if your kitten starts going over you with that rough little tongue, you can assume it is telling you that it considers you a member of its family.

DON'T DROP THE BALL

Once you have established a bond with your kitten, you will need to nourish it consistently over time. Never take your pet's affection for granted. Even if you are still taking care of all of your kitten's tangible needs, if you neglect your pet emotionally, your relationship can suffer. A kitten that has come to expect a play session as soon as you arrive home each day will resent it if you suddenly discontinue the tradition. Sure, some days your kitten won't be in the mood for play, but it should always have the option.

Bonding time doesn't have to be all about recreation. You can also build trust and deepen your relationship with your kitten through other activities, such as grooming. Perhaps you send your longhaired kitten to a groomer for baths. If so, you might consider performing this task yourself instead. If you prefer to leave the more complicated jobs to a professional, consider the ones that require virtually no experience. To your kitten, even simple brushing can feel like a luxurious massage.

Several breeds including the Savannah, Bengal, and Sphynx are known for being highly trainable. The Pixie-bob is even known to enjoy walking on leashes. Mixed-breed kittens are also capable of learning tricks if their owners are willing to invest a little time and effort. How quickly your pet learns or what it can do is far less important than the time you spend together trying. Keep training sessions short and fun, and always reward your kitten for even the smallest steps in the right direction. If you praise your pet profusely for partially meeting a goal today, it is far more likely to keep working toward that goal tomorrow. And if you have deepened your bond in the process, you have both already succeeded.

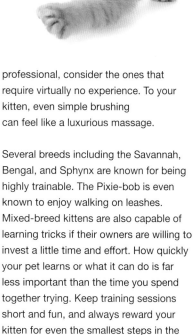

THE GREEN-EYED MONSTER

What should you do if your kitten appears to be bonding more closely with someone else in your household than with you? While it's normal to feel a twinge of jealousy in this situation, resenting the relationship your kitten is building with the other person won't do a thing to deeper your own connection with your pet. Instead, consider what the other person might be doing that you aren't. For instance, is this person feeding the kitten? The feline species often identifies most strongly with the person who delivers its meals. Next, be sure you are making time to play with your kitten every day. This activity can help your pet associate you with enjoyment. Finally, instead of competing with your housemate for your kitten's attention, try imitating the other person's behavior. It could simply be that you are trying too hard while the other person is simply letting the kitten come to him or her.

KNOW THE SYMPTOMS AND SOLUTIONS

ven if you do everything in your power to select a healthy kitten, it may still get sick from time to time. Breeders can—and should—provide a health warranty for their animals, covering specific major illnesses or orthopedic problems. No one, however, can promise you that these problems won't strike. Fortunately, many of the health issues that can affect your young pet are not serious, providing you know how to handle them.

Whether you are dealing with a minor illness, a major disease, or an injury, the most important thing for you to do is remain calm if your kitten becomes sick or hurt. Many owners would rather experience pain themselves than see their precious pets suffer, but remember that you can only help your kitten if you keep your wits about you. Crying or screaming can scare a young cat, adding to any stress the animal is already feeling. By maintaining your composure, you can calm your kitten and get it the help that it needs as quickly and efficiently as possible.

UPPER RESPIRATORY INFECTIONS

Among the most common illnesses that afflict kittens are upper respiratory infections (URIs). Although any cat can suffer from this problem, it is much more pronounced in young cats. Each infection can be a bit different, depending on the specific virus or bacterium that caused it, but the symptoms are usually the same. A kitten fighting an URI may cough or sneeze, experience runny eyes and noses, and move sluggishly. The animal may also understandably have a decreased appetite.

Some people compare feline upper respiratory infections to the common cold in people. Like colds, URIs are most commonly caused by viruses and can be largely unaffected by treatment with antibiotics. Also like the common cold, a feline upper respiratory infection is highly contagious. It can take anywhere between a couple of days and a week and a half for your kitten to show symptoms of the illness once it enters its system. An average infection can then last from one to three weeks. Your pet will be contagious this entire time.

Most upper respiratory infections can be treated at home, but it is smart to contact your veterinarian as soon as you notice the symptoms. Your vet can help identify the intensity of your kitten's case and

prescribe medications, such as eye ointments, if necessary to help speed recovery. At home you may be able to stimulate your pet's appetite by warming wet food. Using a vaporizer can help if your pet is experiencing nasal congestion. If you don't have one, you can achieve a similar effect by taking your kitten into the bathroom, closing the door, and running a hot shower for about 15 minutes.

 Kitten Fact

In the wild a sick cat faces a greater risk of becoming a larger animal's prey. For this reason wild cats are skilled at masking infirmities. The instinct to hide illness remains strong in domestic cat species as well, so it is particularly important that owners are always on the lookout for symptoms of a problem.

Common Kitten Emergencies

If your kitten experiences any of the following problems, seek veterinary care immediately.

🐾 **Allergic reactions**

🐾 **Burns**

🐾 **Choking**

🐾 **Cuts and bleeding**

🐾 **Consuming foreign objects**

🐾 **Difficulty breathing**

🐾 **Electric shock (from cords)**

🐾 **Fever**

🐾 **Poison ingestion**

🐾 **Trauma**

DIARRHEA

Abrupt changes in diet can quickly lead to diarrhea for your kitten, but did you know that stress alone can also cause this common problem? While you may not think that your tiny pet has much stress in its life, consider all the changes that your kitten has experienced recently. If you just recently brought it home, it has left both its mother and its siblings and been placed in a brand-new home it had never before seen—all with no warning whatsoever.

If your kitten is passing loose stools, keep an eye on the situation. If the problem continues for more than a day or two, call your veterinarian. The first thing the vet will want to do is rule out a physical cause, such as an intestinal issue. Diarrhea is a common symptom of worms, for instance. Once a physical cause has been eliminated, you can then work on getting your pet's digestive system back in order.

Whether your pet's diet or its nerves have caused the problem, your vet will probably recommend withholding food for about 24 hours. After giving your kitten's system this short break, you can then start offering small amounts of food gradually. Be sure that your pet continues drinking water throughout this period, as diarrhea can cause dehydration.

BE PREPARED

Although many health issues will require a trip to the veterinarian, you can make virtually any medical crisis go more smoothly by keeping a first-aid kit for your kitten. Stock it with the following items, and make sure that everyone in the household knows its location in case of an emergency.

- Activated charcoal
- Antibiotic ointment
- Antiseptic spray
- Clean towel or blanket
- Corn syrup
- Cotton balls and swabs
- Disposable gloves
- Elizabethan collar
- Flashlight (small penlight is ideal)
- Gauze pads and nonstick bandage rolls
- Hydrocortisone cream
- Hydrogen peroxide
- Instant cold packs
- Magnifying glass
- Rectal thermometer
- Saline solution
- Scissors
- Styptic powder
- Syringe (no needle)
- Tweezers
- Any medications your kitten takes
- Index card with important phone numbers— veterinarian, emergency vet, poison control, etc.

INTESTINAL PARASITES

Speaking of worms, these and other intestinal parasites are also among the most common health issues that strike kittens. Some kittens are born with worms, most commonly hookworms and roundworms. Some breeders make deworming a part of their protocol with each litter; other breeders prefer to treat only the kittens that suffer from this problem. Deworming all kittens certainly helps keep the problem at bay, but critics of this approach insist that animals shouldn't be given unnecessary medications.

If left untreated worms can take a toll on your kitten's body. Signs of roundworms include diarrhea, vomiting, and a lack of appetite. A kitten afflicted by roundworms also may not grow as quickly as it should. Roundworms usually aren't fatal in adult cats, but if too many of these parasites accumulate in your kitten's intestines, the situation can become life threatening. Symptoms of hookworms include hair loss, digested blood in the stool, and weight loss, but the overwhelming sign of this type of worm is the anemia that it causes. A kitten's prognosis in fact depends on how anemic it has become by the time it receives treatment.

The best way to make sure that your kitten isn't suffering from worms is to take regular stool samples to the vet—ideally each time you go in for a routine checkup. If you notice any of the above signs, however, ask your vet to check your pet's stool right away. You should be able to see roundworms in your kitten's feces, as adult worms measure between 3 and 6 inches (8 to 15 centimeters) long. Hookworms, on the other hand, are only about 1/8 of an inch (2 to 3 millimeters) long, so they often go unnoticed by an owner's naked eye. Once your vet confirms a case of worms, they can begin the necessary deworming medication.

 Kitten Fact

A kitten piller can make administering medication to your pet significantly easier. This device, which works a bit like a syringe, helps you drop the pill down your kitten's throat, without having to force your hands into its tiny mouth. To prevent getting scratched in the dosing process, wrap your pet in a blanket before you begin. Once you have expelled the pill, rub your kitten's throat until it swallows.

EXTERNAL PARASITES

If your kitten appears to be scratching constantly, fleas could likely be the cause. Like hookworms, these external parasites can cause anemia in young cats. Fleas can also spread serious illnesses to your kitten through their bites. For these reasons it is important to rid your pet of a flea infestation as quickly as possible. Because of your kitten's age, though, talk to your veterinarian before choosing a product. Certain chemicals that are safe to use on adult cats are too strong for your younger pet.

Depending on the color, length, and thickness of your kitten's coat, you may or may not be able to see the fleas themselves. After combing your pet, however, wipe the comb on a damp paper towel. If you see small specks of dark red or black, you can assume that they are flea dirt—the digested remains of your pet's blood. Even if you find no fleas, a trip to the vet is in order if your kitten is scratching incessantly. No matter what the cause, it is in your pet's best interest to get to the bottom of the problem.

It your kitten is suffering from fleas, you will also need to treat your home. These nasty little creatures can attach themselves to curtains or furniture upholstery, and even delve into your carpets. And they can bite people just as easily as animals. You must remove all the people and pets from your home before treating it with a flea bomb or fogger. Check the package directions to see when it will be safe to return. After you come back, vacuum your home thoroughly, and then toss the bag immediately.

Like fleas, mites cause severe itching, but in the case of these tiny arachnids, the scratching is typically limited to your kitten's ears. You may notice a waxy, brownish-black discharge coming from the ears that looks a bit like coffee grounds. Once your veterinarian diagnoses mites, he or she can then begin treatment with medication. This is a lengthy process—a minimum of three weeks, as no medication can penetrate mite eggs at present. Instead, vets must wait for them to hatch and kill the new mites as they appear.

Whether the culprits are fleas or mites, it is important that you seek veterinary care before your kitten hurts itself with all the scratching. Open sores surely hurt, but they are also a breeding ground for infection. By acting quickly, you may also be able to limit the number of creatures that succeed at making a home in your kitten's coat.

No. 22 YOUR KITTEN AND VET VISITS

FROM VACCINATIONS TO SPAYING/NEUTERING

Visiting the veterinarian with your kitten won't take up much of your time, but the importance of these routine vet visits cannot be underestimated. Some owners are tempted to wait to take a new pet for its first exam until a trip to the vet seems more necessary. They may assume that veterinary care is something that only sick or older animals need. Other owners might worry that taking a young kitten to a veterinary hospital will expose it to contagious illnesses other patients are fighting. If you want your healthy kitten to stay that way, though, the fact of the matter is that preventive veterinary care will play an essential role in the process.

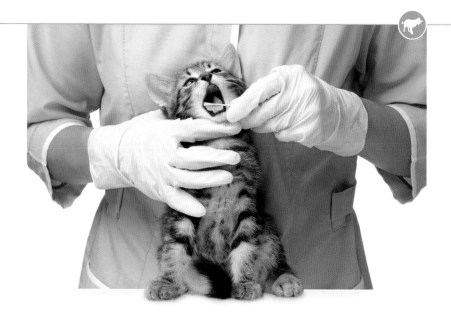

SPOTTING THE SIGNS

Healthy kittens share many obvious qualities, but sick animals can be tougher to spot than you may think. Veterinarians are trained to pick up on the subtlest signs of illness, sometimes long before an owner notices a problem. The biggest advantage you can give your pet for fighting virtually any disease is early diagnosis. Left untreated, many minor problems can morph into more serious ones surprisingly quickly.

If you are concerned that your kitten will pick up a virus or other illness while visiting the vet, take reasonable safety precautions. For starters, always transport your kitten in its carrier, and only remove it from this enclosure for its actual exam. Keeping your pet in its crate can also provide a sense of security if your kitten feels nervous or otherwise distressed. Even inside the examination room, don't allow your pet to jump down onto the floor. The exam table should be sanitized between patients, so this is the safest spot for your kitten.

Some veterinary hospitals offer separate waiting rooms for well visits and sick pets. When interviewing potential vets, consider whether a facility offers this amenity. While this feature certainly shouldn't be the most important item on your pro-con list, it may be worth noting, especially if two hospitals are otherwise equally deserving of your business.

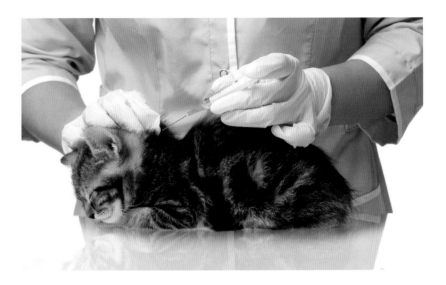

VACCINATIONS

Among the reasons your kitten needs to visit a veterinarian soon after its homecoming is to keep up with the vaccination process. Whether you bought your kitten from a breeder or adopted it from a shelter, an immunization record should be among the paperwork you were given. This document will list all the vaccines that your pet has received along with the specific dates they were given. Some may require booster shots to ensure immunity, but it is important that these additional inoculations be delivered on the correct timetable.

If you plan to allow your kitten to roam outdoors, your pet's risk of catching certain diseases will be higher. Likewise,

if you plan to show your pet or participate in other activities that will expose your kitten to other animals, vaccinating for the most common contagious illnesses may be a smart step. Some organizations may even require that a kitten have certain shots before being granted admittance.

Vaccinations have become a controversial issue in recent years, with many owners worrying that certain vaccinations—or too many shots within a small time frame— can cause other health problems. Some vaccines are mandated by law, but others are a matter of personal choice. Your best resource for information about vaccines and their possible side effects is your veterinarian. Together you can decide which shots your kitten needs.

Kitten Fact

Your kitten may feel a little tired or irritable after receiving a vaccination. This is perfectly normal. If either of these side effects continues for more than a few days—or if your pet appears to be ill—report its symptoms to your veterinarian at once.

Recommended Vaccinations

The American Veterinary Medical Association (AVMA) recommends the following feline vaccinations for healthy animals:

- 🐾 Rabies

- 🐾 Feline Panleukopania Virus (FPV)

- 🐾 Feline Herpesvirus-1

- 🐾 Feline Calicivirus

- 🐾 Feline Leukemia Virus (FeLV)

After the initial vaccination, your kitten will need regular "booster" shots. Your veteriniarian will be able to advise you on the best schedule of treatments for your pet.

THE FIRST VISIT

Your kitten's first veterinary visit should be quick and simple. Your veterinarian will likely begin by taking your pet's temperature. He or she will then conduct a brief physical examination on your kitten, checking its eyes and ears, teeth, coat, and musculature. The exam should also include weighing your pet to make sure it is growing properly. This step affords an excellent opportunity to discuss your kitten's diet. Tell the vet what and how much you are feeding your kitten to see if any changes are necessary.

Just as your veterinarian is a valuable resource about vaccinations, the vet can also offer you a wealth of knowledge about kitten ownership in general. If you are encountering any problems with your pet, share them with this caregiver. Whether the issues relate to health or behavior, your vet may be able to point you toward a helpful solution you might not have considered yet.

Your vet will also ask *you* some questions about your kitten's care, history, and general well-being. Answer each one as thoroughly and honestly as possible. Your veterinarian won't think poorly of you if your pet is having a hard time settling into its new home. Good vets understand that transitions can be difficult on both animals and owners, and they want their patients to lead the best lives possible.

As you prepare for your kitten's first veterinary checkup, you may realize just how many questions you have for the veterinarian. When the time comes, though, you may become too distracted by a nervous or meowing kitten to remember all your questions. To ensure that you don't forget to ask about something important, write down each question or concern that you have as it strikes you. Bring the list with you to the appointment, but wait until the end of the checkup to go over it with your vet. He or she may very well address many of your concerns by the time the exam is finished, but don't be afraid to ask any questions that remain unanswered. A good vet will be impressed that you are concerned enough about your kitten to educate yourself as much as possible about its care. If you are concerned about taking up too much of the vet's time, head online before your appointment to see if the hospital has a website. If it does, as many now do, you may

be able to find some of the information you are seeking there. Websites can also be helpful for obtaining information that you need before an exam, such as when to get and how to transport a stool sample.

Kitten Fact

Feline Leukemia Virus (FeLV) and Feline Immunodeficiency Virus (FIV) are deadly diseases that presently have no treatments. For this reason many veterinarians recommend having a kitten tested for both illnesses during its first checkup.

SPAYING/NEUTERING

According to the American Society for the Prevention of Cruelty to Animals (ASPCA), kittens can be spayed or neutered safely as early as eight weeks of age. Your veterinarian may suggest waiting until your pet weighs at least two pounds (1 kilogram), though. Before you leave the veterinary hospital from your kitten's first checkup, you might want to schedule this important procedure.

When you have your female kitten spayed, you significantly reduce its chance of suffering from mammary cancer—the feline form of breast cancer. Because the operation removes the animal's ovaries and uterus, it completely eliminates the threats of ovarian and uterine cancer. Similarly, having a male kitten neutered prevents it from suffering from testicular cancer.

Many veterinarians and owners alike also point out that sterilization makes better pets. Female cats in the midst of heat tend to yowl incessantly. They also attract a great deal of attention from wandering male cats that can hear and smell a female in season from an impressive distance. Male kittens are much less likely to mark their territory with urine, a relatively common issue in intact male cats.

 Kitten Fact

According to the Winn Feline Foundation, studies have revealed some surprising benefits to spaying and neutering kittens before they enter puberty. One such study found that male kittens sterilized between the ages of six and 16 weeks of age suffered fewer incidents of asthma and gingivitis.

BEHIND THE SIGNS

Parents Don't Make Better Pets

Numerous myths surround the spaying and neutering of animals. Some owners insist that there are health benefits to breeding a female. According to the Humane Society of the United States (HSUS), the opposite is actually true. Female kittens spayed before their first heat cycle tend to be healthier than those that have given birth. Another myth is that a male cat will feel less masculine if he is neutered. Sterilization will not have a negative effect on your male kitten's emotional well-being or his personality. If anything, his temperament may be improved.

MAKE A PLAN

Another appointment you may consider making before you leave your kitten's first checkup is its next routine examination. As long as your pet remains healthy, you won't need to return for this exam for another year. Most veterinarians recommend an annual "well" visit for young cats. If any health issues arise between now and your kitten's next checkup, though, definitely pick up the phone. Even if that next exam is right around the corner, don't wait. Prompt treatment is essential for defeating many health problems.

Most veterinary hospitals send out reminders when a pet is due for an exam or vaccines. If your kitten is due for a vaccination that doesn't match up to its checkup schedule, consider two separate visits instead of one. In addition to spreading out the cost a bit, you will be making sure that your pet is getting everything it needs, exactly when it needs it.

No. 23 KITTEN PLAY

EXERCISING MIND AND BODY

When people get new puppies, exercise and play seem like prerequisites. Everyone knows that dogs need to be walked, for instance, but owners often overlook a kitten's need for activity and fun. Kittens may be little, but boy, do they possess lots of energy! They need physical stimulation to stay fit and mental stimulation to be happy. If they aren't provided with outlets for all their vitality and intelligence, they may even develop behavior problems.

Play also offers kittens an important opportunity for socialization. Unlike a puppy that goes for a daily walk, your kitten may not encounter people and other pets very often. Making a point of playing with your kitten helps strengthen its bond with you, and helps ensure that it sees you and the rest of its human household members as part of its family.

TOY WITH YOUR KITTEN

The most obvious way to play with a kitten is with toys. When it comes to these playthings, you may notice that your kitten has some definite preferences. Your pet's breed or its individual personality can exert great influence on which items it enjoys the most. For example, breeds with the strongest hunting instincts include the Abyssinan, Munchkin, LaPerm, and Siamese. These kittens may appreciate battery-powered mice or feather toys. A favorite toy doesn't have to mimic an actual animal to appeal to these breeds, though. It's the act of hunting that makes toys fun for these highly predatory breeds. Virtually any toy that moves on its own or can be dangled at the end of a string will probably appeal to your little hunter.

If your kitten is especially smart, you may need to look for toys that offer an added challenge. Puzzle toys or balls that release treats when rolled a special way are ideal for smart breeds like the Burmese, Cornish Rex, and Singapura. If a toy dispenses an edible reward, be sure it is something that your kitten likes, though, or it may lose interest in winning the game rather quickly.

Another toy that many kittens enjoy is a ball. Your kitten may simply want to chase this item, but it may also retrieve it. The Abyssinian, Bombay, and Chartreux are notorious for playing fetch with their owners. Whether your kitten merely dashes after a ball or brings it back to you, playing with this toy is sure to provide your pet with an invigorating exercise session.

THINK OUTSIDE THE STEREOTYPE

Cats and kittens often get pigeon-holed as pets that like certain things—and only those things. But playful felines enjoy doing much more than chasing toy mice and batting at yarn. The Manx, for example, is often described as dog-like. Like the Abyssinian, this breed can easily be taught to walk on a leash and enjoys going for walks with its owner. The key to taking advantage of this natural inclination, though, is doing the training while your cat is still a kitten. Behaviors that are reinforced during kittenhood are more likely to continue through adulthood.

Like the Turkish Van, the Manx also enjoys playing with water. If you bring a Manx kitten home during an especially warm time of the year, setting up a small rigid kiddie pool in your backyard could lead to a whole lot of fun for both you and your pet. Always supervise your kitten around water, and keep the level low— just an inch or two is plenty deep enough. Also, don't buy an inflatable pool, as your pet's claws could destroy it. If you don't

want to venture outdoors, you can allow your pet to play in the bathtub instead. If you go this route, leave the water running just a tiny bit. Kittens that enjoy playing in water can hardly resist a dripping faucet. Other breeds that are known for liking water include the American Bobtail, American Shorthair, Bengal, Maine Coon, and Turkish Angora.

In colder months—or if you simply prefer not to have to dry off a wet kitten— consider games that require no props. Maine Coon and Ragdoll kittens have a natural tendency to follow their owners around the house. If you or your pet are feeling playful, you can easily turn this habit into a thrilling game of hide-and-seek. You have a couple of different options for initiating the game. One option is to wait until your kitten is already in pursuit, and then dart into the first hiding spot that presents itself. Chances are good that you won't even have to call to your pet for it to find you, but teasing your kitten with a "Come find meee!" certainly isn't against the rules. The best thing about games like this one is that

you can make up your own rules. You may have to move quickly, though, as your kitten's little legs can already move astoundingly fast.

If you can't seem to escape your pet long enough to hide, you might have to go with another strategy. When your kitten isn't paying attention, settle into your hiding spot, and *then* call to your pet. A game performed this way is also an excellent way to work on teaching your kitten its name. Be sure to reward your kitten for winning the game. Verbal praise or even a tasty treat could make your pet more likely to take part next time.

(paw) Kitten Fact

Kittens that play with water tend to fear baths less as adults.

THINKING BIGGER

Since you can't always be there to play with your kitten when the mood strikes, consider offering your pet an object or two to make independent play and exercise a bit easier. Cat trees can provide kittens with hours of entertainment and exercise opportunities. While navigating these tall, carpeted structures won't provide your pet with the same kind of workout as running across the floor, your kitten will be using various muscle groups as it jumps onto the different platforms and climbs through the openings from one level to another. Playing this way also improves a kitten's balance and coordination. Cat trees are available at most pet supply stores in a wide assortment of sizes and prices.

In addition to exercise, cat trees offer kittens a place to rest, undisturbed by any canine family members. The very top can also be an excellent spot for feeding your kitten if a canine housemate has developed a yen for kitten food. You can encourage your pet to use a tree by placing smaller toys or treats on the structure. Many trees also come equipped with a scratching post, which is good for both claws and leg muscles.

Another fun toy for independent play is a cat tunnel. If you are unsure how much your kitten will enjoy this item, start small. Tunnels measuring only a foot or two in length are extremely economical and should give you an idea of whether a larger setup would be worthwhile. If your kitten darts through the tiny tunnel repeatedly, it might be worth investing in a larger one. You can find tunnels made of nylon or burlap, often with maze-like annexes. Some are even made of see-through mesh with zippers to enclose them for safe use outdoors.

YOUR KITTEN'S BAG OF TRICKS

The debate over which animal is smarter—a cat or a dog—has undoubtedly been raging since the first members of both species were domesticated. Many owners and experts agree that cats are indeed the more intelligent species, yet most owners do not even consider training their feline pets. You may be surprised to learn that one of the first scientific studies about the importance of reinforcement in animal behavior was actually conducted with cats. Kittens in particular are wide open to learning new things. If owners are willing to take the time and make a small effort, they are likely to be amazed by the things their tiny pets can learn.

You can teach a kitten many of the same things you teach a puppy. Your feline pet can learn to sit, jump, and even roll over on command. If these tricks do not interest you, you can also go down a completely different route. Perhaps you would like to teach your pet something more fun, such as shaking hands or walking or jumping through a hoop. By starting small and remaining patient, you can easily teach your kitten any of the above tricks.

As the aforementioned study implies, rewarding your kitten is the key to successful training. Expect mistakes—lots of them—and focus on the successes, however small they may be. Consistency is also important. Your kitten will tolerate frequent yet short training sessions better than longer ones, but you mustn't expect your pet to remember what it learned a week ago unless you have continued to reinforce it with practice, praise, and lots of treats. If you find yourself overfeeding your kitten as a result of training, consider offering your pet one of its meals as its rewards. You may also have the best luck working on trick training when your kitten is hungry, as this is when it is most apt to work the hardest.

GET MOVING

Sir Isaac Newton's first law of motion states that a body at rest stays at rest, while a body in motion stays in motion. If Newton had been talking about certain cat breeds, he would have been right only about the first part. Even the mellowest kittens need exercise and play, but getting breeds such as the Himalayan, Persian, and Ragamuffin moving can be a bit of a challenge. Even young cats of these breeds delight in stationary pastimes. This doesn't mean that your pet *won't* exercise. It simply means that you will probably need to be the instigator when it comes to this part of your cat's healthy routine. The key to exercising a lazy kitten is finding its currency. Perhaps your kitten loves fishing pole toys, or maybe it prefers a laser pointer. Whatever grabs your kitten, exploit it. You may even make a point of keeping these toys put away until your daily exercise time comes to keep the demand high.

No. 24 YOUR KITTEN IS GROWING UP

RECOGNIZE THE SIGNS AND KNOW WHAT TO EXPECT

Every kitten is a little different. Some breeds mature more quickly than others—either mentally, physically, or both. For most breeds, though, young adulthood comes around 18 months of age. What most breeds have in common is that kittenhood seems to pass by in a blur, at least for owners. In the beginning, owners are focused on doing all the right things to make their pet's first weeks and months the best they can be. Buying toys, choosing the best food, and settling their feline friends in their new home keep owners mighty busy. At first it may seem like litter-box training alone will never end. Soon, though, the kitten that used to dash from one end of your home to the other at midnight in a total frenzy becomes a well-adjusted (and reliably housetrained) young adult.

Before you even realize it, your kitten isn't a kitten anymore. As sudden as it may seem, however, adulthood doesn't happen overnight. Instead, your kitten gradually moves from impish youngster to independent grownup. You may be worried that your relationship with your kitten will change as it matures—that it might need you less. While it's true that some things will change, the bond you have begun making with your beloved feline should only strengthen as it gets older. And rest assured, your precious pet still needs you.

SIZING THINGS UP

The first and most obvious sign of early adulthood is increased size. While the number on the scale can vary by the breed, a good way to keep track of your kitten's growth is to weigh your pet regularly. Don't just focus on the end result. While growth spurts can certainly happen, your kitten should gain a little at a time over the first year and a half. If you notice a marked, sudden increase or several weeks without any increase at all, talk with your veterinarian. It may be necessary to adjust your kitten's food or portions, or a health problem could be to blame.

At first your kitten's growth may leave it looking a bit like an awkward teenager—lanky or disproportionate. It may even act a little clumsy as it moves through this normal phase and gets used to its longer legs and increasing strength. Eventually, your kitten will grow into its bigger body, gaining a more balanced appearance and dependable coordination. You may even find your heart in your throat as you watch your young cat jump distances approaching five times its own height from a standing position. Clearly, those muscles your pet has developed are working.

Male kittens usually end up a bit bigger than their female counterparts. When reading through your kitten's breed standard, look for its projected weight. If the standard offers a range, a female will typically fall on the lighter side, with the male on the heavier side of the span. If you don't know your rescued kitten's lineage, it can be tough to know exactly how big it will get. Your veterinarian may be able to help you determine which breeds may be behind your mixed breed, however, which can then offer you at least some idea of its possible adult size.

If a mere guess isn't good enough, consider having a DNA test performed on your cat to identify its ancestry. While this may seem like an indulgence, testing also offers some practical applications. For example, by identifying your cat's unique mix of breeds, you will be able to keep an eye out for the illnesses for which those breeds are most prone as your kitten moves into adulthood. Currently, these tests are approximately 90 percent accurate.

A CALMER, GENTLER PERSONALITY—USUALLY

Both your cat's genetics and its experiences will play roles in its developing personality. Certain breeds are known for being more affectionate than others. The Persian, for instance, has a reputation for being a people pleaser and a lap sitter. The Siamese, on the other hand, may be loving, but it is also notoriously demanding and talkative. Still, *your* Persian or Siamese kitten may not match this information perfectly.

As many kittens move toward adulthood, though, they become calmer than they were as younger cats. This may be your kitten's maturity kicking in, or it might be that your pet has simply run out of trouble to get into, at least for the time being.

Sometimes a kitten's personality can change dramatically upon entering early adulthood. A kitten that used to cuddle may become more standoffish, and a kitten that spent its entire kittenhood running from you might decide that getting petted isn't so bad after all. Whether the turnaround brings a welcome change or not, chances are good that it won't be the last one your pet goes through before it settles into its more permanent, adult habits.

 Kitten Fact

As your kitten comes of age, you can try offering it catnip. While most young cats have no reaction to the plant at all, many older cats delight in sniffing it. Still, only 50 to 70 percent of adults receive a "high" from doing so. Your older kitten may or may not be among this group.

DIET AND EXERCISE

As your kitten enters early adulthood its energy level decreases, while its corresponding metabolism begins to slow down. These changes make it necessary for you to cut back on your kitten's caloric intake, as well as adjust the ratio of certain nutrients in its diet. Young kittens need high-calorie food rich in energy sources like protein and fat. Some kitten formulas are also high in carbohydrates, as they too provide young pets with plenty of energy. Young adults, on the other hand, can start gaining weight if they remain on kitten formulas too long.

You may not even notice the weight gain at first. In many cases the first time owners realize that their young cats are slightly overweight is when they take them for their annual checkup around the age of two. By this time it may be necessary to place your pet on a weight-loss formula. Unless your cat has gained an excessive amount of weight, the diet food should do the trick, especially if you combine it with increased exercise. A better approach, however, is to avoid the problem entirely by adjusting your pet's diet as it leaves kittenhood.

Exercise is another area of concern. Although your pet's energy level won't be quite as intense as it was just a short time ago, your kitten will still need regular activity, and a fair amount of it. One of the biggest mistakes many owners make as their kittens transition into adulthood is thinking that intentional exercise no longer needs to be a priority. While a young adult cat will continue to romp and play on its own, as its owner, you have a continued responsibility to make sure your pet is getting sufficient physical and mental stimulation.

Your maturing kitten may become bored with some of its old playthings. Remember to keep rotating your pet's toys, and consider trying new games to keep playtime interesting. For example, a ball that once captured your kitten's attention when you rolled it across the floor may no longer impress your more mature pet. Instead, try using the ball to play a thrilling game of mock-hockey. Large, shallow boxes work great for this purpose, or you can simply toss the ball into the bathtub. In either case whenever your pet bats the ball, it will bounce off the sides, making chasing after it much more fun.

Life with your older kitten is sure to hold many more changes, but before long your pet will settle into its new adult skin. Saying goodbye to kittenhood can be bittersweet for owners, but as your pet enters its young adult years, be proud of the work you have put into raising your beloved kitten—and the amazing animal your cat is still becoming.

STICK WITH WHAT WORKS, LEAVE BEHIND WHAT DOESN'T

If your kitten has been happy and healthy eating its kitten formula, stick with the same brand and protein source as you transition your pet over to adult food. Your young cat's stomach is much less likely to be upset by this type of change, as it is often a new protein that triggers upset with a new food. If, however, you have not been satisfied with your pet's kitten food, the move to an adult diet is the perfect time to swap to a formula that works better for your cat. Give your young cat some time to get used to the new food, but you needn't stick with it if it too falls short of your expectations. Keep searching until you find just the right option.

GLOSSARY

ANEMIA
A condition in which the blood has less than the normal amount of red blood cells, hemoglobin, or total volume and which is usually marked by weakness, exhaustion, shortness of breath, and abnormal heartbeat.

ANTHROZOOLOGIST
A person who studies the interactions between humans and other animals.

BUNTING
A behavior of cats in which they rub or push their face against people or objects.

CATTERY
A place where cats are kept and bred.

CONFIRMATION
Cat breed shows that evaluate individual animals
in relation to how well they match their breed standards.

CREPUSCULAR
Active during twilight.

DAM
The female parent, especially of a domestic animal.

DANDER
Minute scales from hair, feathers, or skin that may cause allergy.

DOMESTICATE
To adapt to living with human beings and to serving their purposes.

DOMINANCE
High status in a social group, usually acquired as the result of aggression, which involves the tendency to take priority in access to limited resources, such as food, mates, or space.

EPIDERMIS
The thin outer layer of the animal body that in vertebrates forms an insensitive covering over the dermis.

GINGIVITIS
Inflammation of the gums.

HEMORRHAGE
A great loss of blood from the blood vessels especially when caused by injury.

HIP DYSPLASIA
Abnormal growth or development of the hip bones.

HYBRID
The offspring of two animals of different breeds, especially as produced through human manipulation for specific genetic characteristics.

HYPERPARATHYROIDISM
Overactivity of the parathyroid gland.

HYPOALLERGENIC
Less likely to produce an allergic response, as by containing relatively few or no potentially irritating substances.

INBREEDING
The interbreeding of closely related individuals especially to preserve desirable characteristics and eliminate undesirable characteristics from a stock.

KNEAD
To push in and pull out front paws in an alternating motion.

LACTOSE
A sugar present in milk that breaks down to give glucose and galactose and on fermentation gives especially lactic acid. Also milk sugar.

MANGE
Any of several contagious skin diseases of domestic animals and sometimes human beings that are marked especially by itching and loss of hair and are caused by tiny mites.

METABOLISM
The sum of the physical and chemical processes in an organism by which its material substance is produced, maintained, and destroyed, and by which energy is made available.

MARKING
The act of spraying urine on an object as a means of showing ownership.

MOCK-AGGRESSION
A simulation of an offensive action carried out in play and/or for the purpose of learning.

NEUTER
To castrate.

PAPILLAE
A small bodily structure (as one on the surface of the tongue that often contains taste buds) that resembles a tiny nipple in form.

PARASITE
A living thing that lives in or on another living thing in parasitism.

PEDIGREE
A table or list showing the line of ancestors of a person or animal.

PHEROMONE
A chemical substance (as a scent) that is produced by an animal and serves as a signal to other individuals of the same species to engage in some kind of behavior (such as mating).

QUARANTINE
A limiting or forbidding of movements of persons or animals that is designed to prevent the spread of disease.

QUICK
The pinkish region of the nail that includes blood vessels.

SEPARATION ANXIETY
The stress and anxiousness in an individual brought on by the leaving (perceived or imagined) of another individual.

SIRE
The male parent of an animal and especially of a domestic animal.

SOCIALIZATION
A continuing process whereby an animal learns the behavior and social skills necessary to interact with people and animals.

SPAY
To remove the ovaries (of an animal).

STEATITIS
An inflammation of adipose (fatty) tissue.

STYPTIC
A substance designed to speed the clotting of blood.

TAURINE
A sulfur-containing amino acid important in the metabolism of fats.

TOCOPHEROL
A vitamin-based preservative.

VOMERONASAL ORGAN
An organ of chemoreception that is part of the olfactory system of amphibians, reptiles, and certain mammals.

PICTURE CREDITS

79 © Yuliya Ozeran | Shutterstock
80 © Ermolaev Alexander | Shutterstock
81 © AnnaIA | Shutterstock
82 Right © Ermolaev Alexander | Shutterstock
83 © MaraZe | Shutterstock
84 Right © Ermolaev Alexander | Shutterstock
85 © gengirl | Shutterstock
86 Top © Paisit Teeraphatsakool | Shutterstock
86 Bottom © Tsekhmister | Shutterstock
88 Left © Eric Isselee | Shutterstock
88 Top right © Tony Campbell | Shutterstock
89 © Irina Burakova | Shutterstock
90 Left © Elena Efimova | Shutterstock
92 © Eric Isselee | Shutterstock
93 Bottom © DenisNata | Shutterstock

94 Bottom left © Beauty photographer | Shutterstock
94 Bottom right © Jagodka | Shutterstock
95 © Schubbel | Shutterstock
96 © Oksana Kuzmina | Shutterstock
97 Left © Stefano Garau | Shutterstock
97 Bottom right © cath5 | Shutterstock
98 © Ermolaev Alexander | Shutterstock
99 © foaloce | Shutterstock
100 © otsphoto | Shutterstock
101 © sad444 | Shutterstock
102 © attila dudas | Shutterstock
103 © otsphoto | Shutterstock
104 Left © nvelichko | Shutterstock
104 Right © nelik | Shutterstock
105 Left © Sonsedska Yuliia | Shutterstock
105 Right © Eric Isselee | Shutterstock
106 © Jagodka | Shutterstock
108 © Wallenrock | Shutterstock
109 © Offscreen | Shutterstock

BIBLIOGRAPHY

American Association of Feline Practitioners—www.catvets.com/public/PDFs/PracticeGuidelines/FelineBehaviorGLS.pdf

American Veterinary Medical Association (AVMA)—www.avma.org/Pages/home.aspx

American Society for the Prevention and Cruelty of Animals (ASPCA)—www.aspca.org

Andreassia, Katia. *"New Documentary Condemns Declawing of Cats; Who is Right?* National Geographic Daily News, September 4, 2013—news.nationalgeographic.com/news/2013/09/130911-declawing-cats-paw-project-vets-pets

Animal Behavior Society—www.animalbehavior.org/ABSAppliedBehavior/article-applied-behavior/why-cats-scratch-things

Animal Planet—www.animalplanet.com/pets/10-tips-for-bringing-a-new-kitten-home.htm

Animal World—cats.animal-world.com/Hybrid-Cats/Hybrid-Cat-breeds.php

Association of Animal Behaviorist Professionals—www.associationofanimalbehaviorprofessionals.com/let_us_prey.html

Arrowsmith, Claire. *The Kitten Pack: Making the Most of Kitty's First Year*. Irvine, California: Bow Tie Press, 2009.

Becker, Marty. *Your Cat: The Owner's Manual*. New York, New York: Grand Central Life & Style, 2012.

Cat Behavior Associates—www.catbehaviorassociates.com/pica

CatChannel.com—www.catchannel.com/kittens/health/article0009.aspx

The Cat Fanciers' Association—www.cfainc.org

Cat Training—www.cat-training.co.uk/index.htm

Cat World—www.cat-world.com.au/bringing-your-new-kitten-home

Cats of Australia—www.catsofaustralia.com/aggressive-kitten.htm

Cats Protection—www.cats.org.uk

Catster—www.catster.com

Cinotto, Laurie. *The Itty Bitty Kitty Committee: The Ultimate Guide to All Things Kitten*. New York, New York: Roaring Brook Press, 2014.

Companions Animal Hospital—www.companionsanimal.com/boisevet-discusseshestdiet-cats

Cornell University College of Veterinary Medicine—www.vet.cornell.edu/FHC/health_resources/Vaccines.cfm

Galaxy, Jackson. *Cat Daddy*. New York, New York: Penguin, 2012.

Goathouse Refuge—www.goathouserefuge.org

HealthyPets—healthypets.mercola.com/sites/healthypets/archive/2010/06/02/should-you-ever-try-to-adopt-kittens-you-find-in-the-wild.aspx

Humane Society of the United States—www.humanesociety.org

The International Cat Association—www.tica.org/index.php

Live Science—www.livescience.com/24174-high-protein-cat-diet.html

Metzger Animal Hospital—metzgeranimal.com/news/the-midnight-crazies

Oxford-Lafayette Humane Society—www.oxfordpets.com/index.php?option=com_content&view=article&id=61

Paw Nation—www.pawnation.com

Paws on Your Heart—pawsonyourheart.com/facts-about-cats-why-do-cats-have-scratchy-tongues

PDSA—www.pdsa.org.uk/pet-health-advice/kittens-and-cats/pedigree-health

Perfect Paws—www.perfectpaws.com/help3.html#.U49irfldX4Z

Pet Finder—www.petfinder.com/cats/cat-behavior-and-training/how-to-teach-a-cat-tricks

Pet Happy—www.pet-happy.com/at-what-age-can-a-kitten-leave-its-mother-and-littermates

Pets Web MD—pets.webmd.com

Preidt, Robert. "Cat's Genome Map Purrfected." ABC News—abcnews.go.com/Health/Healthday/story?id=4509233

Prynne, Miranda. "RSPCA 'full to capacity' due to cat population crisis." The Telegraph, April 3, 2014—www.telegraph.co.uk/lifestyle/pets/10740363/RSPCA-full-to-capacity-due-to-cat-population-crisis.html

Royal Society of Prevention of Cruelty to Animals (RSPCA)—www.rspca.org.uk/ImageLocator/LocateAsset?asset=document&assetId=1232734779317&mode=prd

Santa Barbara Humane Society—sbhumanesociety.org/services/behavior/kitten_problems.php

Siegal, Mordecai. I Just Got a Kitten. What Do I Do? New York, New York: Fireside, 2009.

Tayman, David. "Howard Pets: Whose nose is keener, cat or dog?" Baltimore Sun, April 1, 2011—articles.baltimoresun.com/2011-04-01/explore/bs-exho-howard-pets-whose-nose-is-keener-cat-or-dog-20110509_1_fish-surplus-food-nose

UC Davis Veterinary Medicine—www.vgl.ucdavis.edu/services/cat/ancestry/faq.php

Universities Federation for Animal Welfare—www.ufaw.org.uk/documents/CALVER.pdf

Vancouver Orphan Kitten Rescue—www.orphankittenrescue.com/urinary-crystal-in-cats-why-not-to-feed-only-dry

VCA Animal Hospitals—www.vcahospitals.com/main

Vet Street—www.vetstreet.com

Virginia-Maryland Regional College of Veterinary Medicine—www.vetmed.vt.edu/vth/sa/clin/cp_handouts/Nutrition_Growing_Kitten.pdf

Web Vet—www.webvet.com/main

Winn Feline Foundation—/www.winnfelinehealth.org/pages/feline_hip_dysplasia_web.pdf

INDEX